HUMAN HEARING
AND THE REALITY OF MUSIC

ARMIN J. HUSEMANN

Human Hearing and the Reality of Music

SteinerBooks

2013

Acknowledgments:
Thanks to Thieme-Verlag for permission to reprint Figures 2, 8, 9, 10, and 11 and to Springer-Verlag for Figures 4 and 14.

Copyright © 2013 by Armin J. Husemann

SteinerBooks
Anthroposophic Press

610 Main Street
Great Barrington, Massachusetts 01230
www.steinerbooks.org

Library of Congress Cataloging-in-Publication Data is available on request.

ISBN: 978-1-62148-048-8
eBook ISBN: 978-62148-023-5

All rights reserved. No part of this book may be reproduced in any form without prior written permission from the publisher, except for brief quotations embodied in critical articles for review.

Contents

Foreword xi

CHAPTER 1
Human Hearing and the Reality of Music 1

The Eardrum   1
Gravity-free Limbs in the Ear and Eye   2
Hearing, Chewing, and Swallowing   4
The Evolution of Hearing: from Earth Vibrations to Air Vibrations   5
The Function of the Inner Ear   7
From Technically Conceived Physiology to Living Physiology   12
Walking, Singing, Hearing   13
A Symbolic Experiment   16
The Entire Muscular System "Hears"   17
Music and the Mirror Neurons   17
Music Therapy   18
The Eye Is Destructive; the Ear Is Motivating   19
The Proof of the Snake   20
The Movement of Music from Note to Note   21
The Reality of Music   22
Karmic Development and Eversion   23
The Everted Will in Hearing and the Ear   24
The Inner Ear and the Intestine   26
The Musical Origin of the Digestive System   28

CHAPTER 2
Music Is "Chemistry from the Inside"
The Fluorine Process in the Human Body   30

The Fluorine Process from the Chemical Perspective   35
The Seven Stages of Life   36
The Genesis of the Ethers and the States of Matter   36

The Number Ether 38
The Periodic Table as a "Creation Document" 39
The Melody of Life and the Actions in the Outer World 40
A Second Human Being Emerges 42
Finger Dexterity and Dental Health 46
Fluoricum Acidum for Severe Insomnia Due to Restless Leg Syndrome 48
Patient History and Diagnostic Findings 49
Finding a Treatment 50

CHAPTER 3
The Experience of Music and Its Basis in Physiology 52

Experiencing Time in Music 52
The Archetypal Music of Singing 56
What Are the Physiological Foundations of Musical Perception of Time? 56
History of Research into the Respiratory Dynamics of Cerebrospinal Fluid 58
How Does Respiration Move the Cerebrospinal Fluid? 59
Abdominal Veins as Mediators between Respiration and Cerebrospinal fluid 62
The Extension of Respiration into the Interior of the Cerebral Ventricles 62
Respiratory Dynamics, the Inner Ear, and the Experience of Music 64
Breathing between Life and Consciousness 65
How the Cerebrospinal Fluid Moves with Respiration 65
An Imaginative Physiology of the Brain's Respiratory Dynamic 67
Respiration as the Organ of Creative Thinking 68
Breathing Colors and Sounds 71
Legato and Staccato—Articulation and Phrasing through Breathing 72
Dionysus, or Breathing Downward 76
Inhalation and Musical Inspiration 79
The Hearing Process in Inhalation 79
CSF-Contacting Neurons 80
Apollo's Lyre—from Chemical Process through Respiration to Light 82
The Liver, on Listening to Symphonies 85
Lunar Forces in the Human Body and in Music 86
Respiration between Sun and Earth 90

Afterword
The Unique Status of Central Europe's "Classical" Music   94

Vienna, the Earthly Home of Classical Music   96

Glossary   98

References   105

Foreword

Music is more than simply something we *hear*. We experience it and love it; it is a primal human need. If, as the pianist Alfred Brendel put it, we are able to "take music at its word,"[1] we confront the question that has also moved this author since his adolescent years: "What takes hold of me when I experience music?"—What reality touches me when music is playing? Another question follows: "What happens physiologically in the human body when we make music and experience music? I will approach these questions from three perspectives: first, from my personal experience and active love of music; second, from the physiological perspective; and third, from the perspective of Rudolf Steiner's spiritual science.

The first chapter deals with hearing, a sensory activity that encompasses far more than processes in the ear and in the brain. The whole body is involved in hearing. In this chapter, we will grapple with the dynamics of will involved in music—dynamics that originate in the metabolism and limbs. Schopenhauer's and Rudolf Steiner's understanding of music as "will illumined by deliberate consideration" becomes the gateway to the spiritual reality of music. An initial incomplete draft of this chapter has already been published in essay form; here, the language has been reworked and the content expanded at important points. As noted in the text, some of the basic ideas presented here come from my father, Gisbert Husemann.

The second chapter deals with the intimate relationship between music and chemistry. Music is "chemistry from the inside." An incomplete version of this chapter was also first published in essay form. Having experienced these ideas in clinical instruction, a colleague and student at the Eugen Kolisko Academy discovered a treatment that was able to cure a patient's symptoms (see pp. 48-51). Her discovery confirmed music as a means of accessing the chemical effects of substances.

Only while working on the final version of this book, however, did I became aware of how closely the musical path (which seeks to enter the chemical ether via the sound ether) relates to the legitimate alchemical efforts of the Rosicrucians. Goethe's advertisement for his *Elective Affinities* (see p. 38) alerted me to this connection.

The third chapter, published here for the first time, deals with the musical experience itself and its physiological foundations. All musicians, even those who do not sing or play wind instruments, know they breathe with the music as it is happening. To date, the significance of this fact has not been the focus of much research on the neurophysiology of musicians. From 1915 to 1918, Rudolf Steiner developed a physiology of artistic imagination based on the movement of cerebrospinal fluid during respiration. This book is a first attempt to relate Steiner's research to the findings of natural-scientific research. We owe our knowledge of cerebrospinal fluid (CSF)-contacting neurons in the cerebral ventricles largely to the work of researchers B. Vigh and I. Vigh-Teichmann of Budapest, but the function of these neurons, which remains a puzzle to medical science, makes sense only in the context of Steiner's research. The same is true of the findings of Ernst Pöppel of Munich, who studies the neurophysiological basis of *time consciousness* and how humans grasp temporal structures—a central issue in music. Pöppel's research tells us that the functioning of the brain's "integration system" is rhythmical, with patterns that repeat approximately every three seconds. In view of Rudolf Steiner's physiology of artistic imagination, this pattern points to the respiratory rhythm reflected in cerebrospinal fluid, with its eighteen repetitions per minute. To sum up briefly, brain function depends on respiration, which is the physiological basis of emotional and creative intelligence. Feeling is based on respiration, not brain metabolism.

Errors are unavoidable in studies that explore new scientific territory, and I will be grateful to readers who point out any such errors in this work. Readers who are interested in the questions of methodology that arise in the presentations in this book should read the lectures of January 3 and 6, 1923, in Rudolf Steiner's *The Origins of Natural Science* (CW 326). In these lectures it is pointed out that a new physiology is a prerequisite for the development of therapy in anthroposophic medicine.

My sincere thanks go to the following colleagues, who read portions of the manuscript and supplied either corrections or confirmation: Phillip Busche, Peter Heusser, Godhard Husemann, Michael Kurtz, Klaus Dieter Schubert, Georg Soldner, Markus Sommer, and Jan Vagedes. In the late 1980s, Friedrich Edelhäuser of the neurophysiological working group of the Medical Section of the Free School of Spiritual Science (Goetheanum) pointed me in the direction of literature that launched the research that led to the third chapter. Sincere thanks to Martina Waldschütz for her work on the manuscript and to Jean-Claude Lin and Thomas Neuerer of Verlag Freies Geistesleben for their close collaboration.

This book would never have happened without the support of the members and friends of the Eugen Kolisko Academy, and the sponsors of this training center for anthroposophic medicine; in particular the Mahle Foundation and the Software AG Foundation. My heartfelt thanks to all of you!

Michaelmas 2010 *Armin J. Husemann*

CHAPTER 1

Human Hearing and the Reality of Music

> *There must exist a knowledge that searches the individual sciences for the elements that will lead human beings back to full life again.*
> RUDOLF STEINER, *The Philosophy of Spiritual Activity*[3]

Anything that becomes audible in our surroundings—footsteps in the next room, a conversation, a car out on the street, the crackling of the heating system, the singing of a bird—all of these audible events have one thing in common: they are produced by movement.[4] When we hear, we participate in movements in our surroundings. This fact prompts us to investigate the ear's integration into the human body from the perspective of movement. As we do so, the course of our investigation will proceed in the same direction as sound itself, from the outside in.

The Eardrum (tympanum)

During the embryonic period, the eardrum develops through an invagination of the skin that is matched by an invagination of the stomodium (oral portion) of the primitive gut. This second invagination extends diagonally backward from the nasopharynx and develops into the Eustachian tube, which will ultimately ventilate the middle ear in the fully developed body (see fig. 1).

The invagination that comes from the outside (skin) and originates in the ectoderm, and the one that comes from within (primitive gut) and originates in the endoderm then meet, forming the outer and inner layers

2   HUMAN HEARING AND THE REALITY OF MUSIC

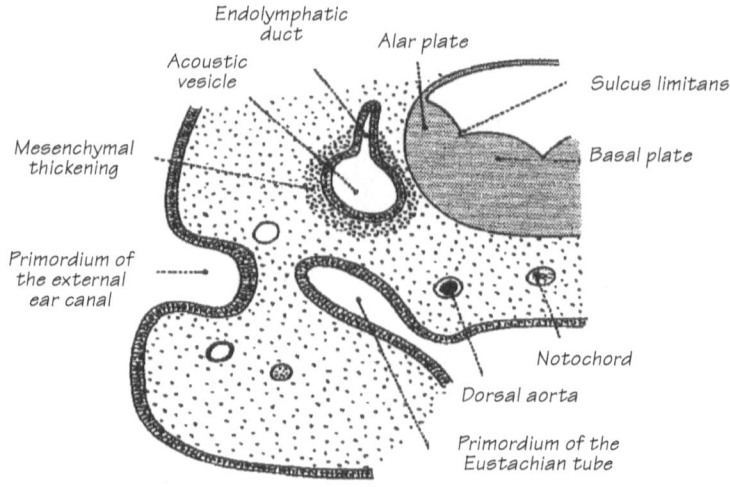

Fig. 1: The development of the eardrum (from W. J. Hamilton, J. D. Boyd, H. W. Mossman, *Human Embryology*, London 1946).

of the eardrum, respectively. Blood vessels, presumably originating in the mesoderm, then grow in between the two layers.

Reddening of the eardrum during a middle-ear infection (otitis media) indicates that the eardrum is well supplied with blood vessels. In the eye, blood is restricted to the area behind the retina because the presence of blood vessels would make the cornea opaque. In the ear, however, sound that enters the ear canal immediately sets the blood vibrating along with the eardrum.

Gravity-free Limbs in the Ear and Eye

The sound that sets the eardrum vibrating is received in the middle ear by three little bones ("hammer," "anvil," and "stirrup") that are connected by real joints to form a "limb" that is moved by sound (fig. 2). Sound vibrations move through both ears into these two tiny limbs. Their joints, like any joints in larger limbs, can be affected by inflammatory joint disease or age-related sclerotic stiffening. When present in the

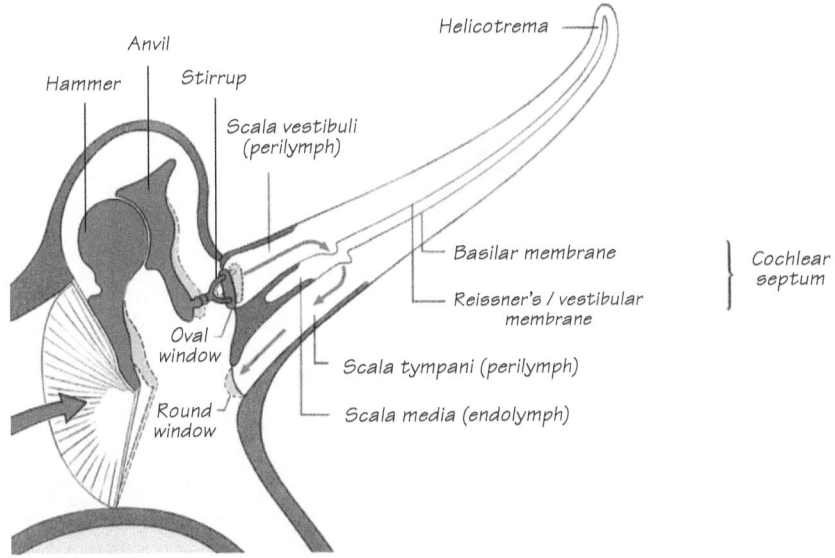

Fig. 2: Sound-induced displacements in the middle and inner ear (according to Zenner).

major limbs, these conditions result in restricted movement in walking or using the hands; in the tiny limbs of the ears, they cause hearing loss. In the major limbs, movement is produced by muscles; in the ossicles of the middle ear, the muscles (the tensor tympani and stapedius muscles) have only regulatory functions. Sound itself takes the place of muscles in moving the bones in the joints. In other words, the transformed limb skeleton of the ear is integrated into sound vibration in the same way that the larger limbs are integrated into musculature. In their movements, the "ear limbs" are totally unaffected by gravity. These auditory bones (called ossicles) vibrate free of gravity because they are suspended in such a way that their vibration produces movement only around their centers of mass.[5] The same is true of the gravity-free "work" of the external eye muscles, which rotate the eyeball around its center of mass.[6]

An older, phenomenologically-oriented way of speaking called light, sound, warmth, and so forth "imponderables"—that is, entities without weight. As described here, the structure of the ear and eye demonstrate the aptness of this term, which Rudolf Steiner reintroduced.

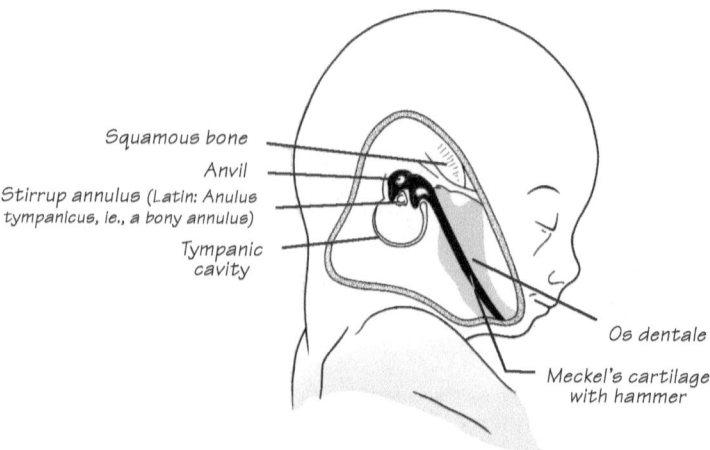

Fig. 3: Semi-schematic depiction of the primary (hammer-anvil) and secondary (squamoso-dental) TMJ joints in a human embryo with a crown-rump length of 62 mm (according to Stark 1979).

Hearing, Chewing, Swallowing

The middle ear cavity, in which the auditory ossicles are suspended, is lined with respiratory mucosa. The middle ear is ventilated through the above-mentioned Eustachian tube, which connects the middle ear and the nasopharynx. We all know the sensation that develops when this ventilation fails; for instance, when the inner opening of the tube is swollen shut because of a sore throat. Air is absorbed into the blood through the mucosa of the middle ear, and a partial vacuum develops when air cannot be replenished from the oral cavity. In healthy individuals, the air supply is replenished with every act of *swallowing*, which automatically opens the Eustachian tube. When we eat, therefore, we "swallow" air into the middle ear.

But the process of hearing also has a deep connection to *chewing*, as the embryonic development of the middle ear demonstrates. Amazingly, during the embryonic and fetal periods in humans, a cartilaginous extension of the lower jaw reaches all the way into the middle ear, where it ends in the hammer, forming a temporomandibular joint (TMJ) with the anvil. In fact, this primary TMJ later becomes the joint

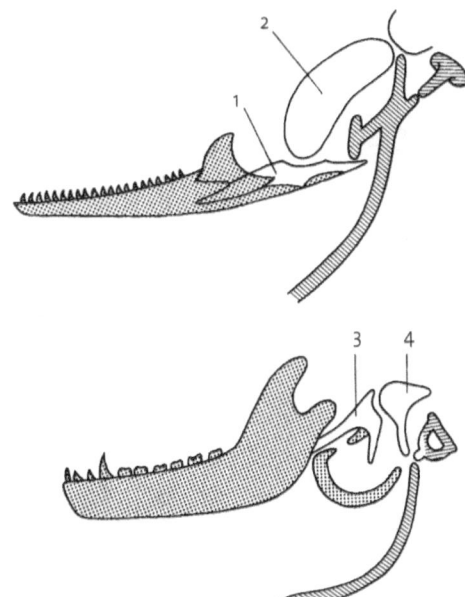

Fig. 4: Metamorphosis of the reptilian temporomandibular joint (1, 2) into the mammalian hammer-anvil joint (3, 4) (according to Gaupp 1911).

between hammer and anvil! In the sixth month of gestation, the lower jaw cartilage (Meckel's cartilage) degenerates, leaving behind a remnant (the hammer), which becomes ossified like the anvil. The joint between hammer and anvil also persists but now serves the hearing process. For chewing, it is replaced by the development of the so-called secondary TMJ. This astonishing metamorphosis of the primary TMJ into a joint that serves hearing (figs. 3 and 4) is the recapitulation of a phylogenetic step between the reptiles and the mammals (Reichert-Gaupp theory of the metamorphosis of the TMJ).[7] Gisbert Husemann sums up this metamorphosis in the following statement: "Hearing is internalized chewing."[8]

The Evolution of Hearing: from Earth Vibration to Air Vibration

The prostrate torso of amphibians and reptiles conducts ground-borne *very low frequencies* to the inner ear via the cerebrospinal fluid (CSF)

6 HUMAN HEARING AND THE REALITY OF MUSIC

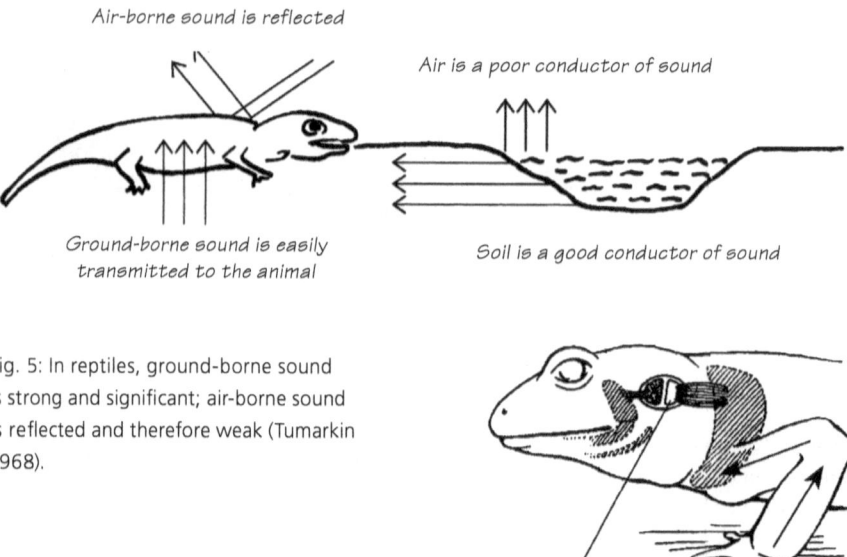

Fig. 5: In reptiles, ground-borne sound is strong and significant; air-borne sound is reflected and therefore weak (Tumarkin 1968).

Fig. 6: How the forelimbs serve hearing (according to Portmann 1976).

system, which is relatively voluminous in these animal classes. This conduction mechanism transmits vibrations and the lowest bass frequencies[9] (fig. 5).

Somewhat *higher frequencies* are transmitted via the forelimbs. For this purpose, amphibians and reptiles have a two-part oval window. The front half is exposed to air via the columella (an auditory ossicle of the middle ear, which corresponds to the stirrup in mammals) and transmits air-borne sound; the rear half is underlaid by a bony plate called the operculum. The ligaments and muscles connecting the operculum to the shoulder blade permit conduction of higher-frequency ground-borne and water-borne sound via the forelimbs to the middle and inner ear.[10] The large size of the vibrating mass of the forelimbs limits hearing of this type to low to mid frequencies (fig. 6).

In mammals, the limbs alternate beneath the torso to support it off the ground. In a parallel development, hearing is internalized: The joint that originally served chewing now serves hearing, and the connection between the eardrum and the forelimbs degenerates. With this step to exclusive perception of air-borne sound, the frequency of perceivable sound rises again.[11] In evolution, perceivable frequencies rise with each decrease in the vibrating mass that transmits them:

↑ Vertebrate evolution	*Auditory ossicles*	high frequencies, up to and including ultrasound
	Forelimbs	low frequencies (<1000 Hz)
	Torso	vibrations

As hearing is internalized, therefore, it also gains access to higher frequencies, and as the frequency of perceived sound increases, so too does the speed of locomotion—the animals begin to walk, then to run, or even to fly. As we shall see, this is no mere outer analogy but the expression of an inherently lawful connection.

In English we speak of "high" and "low" pitches, which correlate with the body's position relative to the earth. It appears to be no coincidence that birds, which rise highest above the earth, also produce the most high-pitched sounds. The validity of instinctively describing sound in spatial terms such as "high" and "low" is also supported by the recent discovery that the roughly four percent of humanity with "amusia" not only cannot perceive differences in pitch within a melody but also have distinct deficits in processing *spatial* forms.[12]

The Function of the Inner Ear

The inner ear is inaccessible to ordinary sense perception. Its scale is microscopic, and it occupies a space—the interior of the bony base of the skull—that typically houses bone marrow, where blood cells are produced. The two-and-a-half turns of its membranous spiral canal lie encased in the spiral channel in the petrous bone.

8 HUMAN HEARING AND THE REALITY OF MUSIC

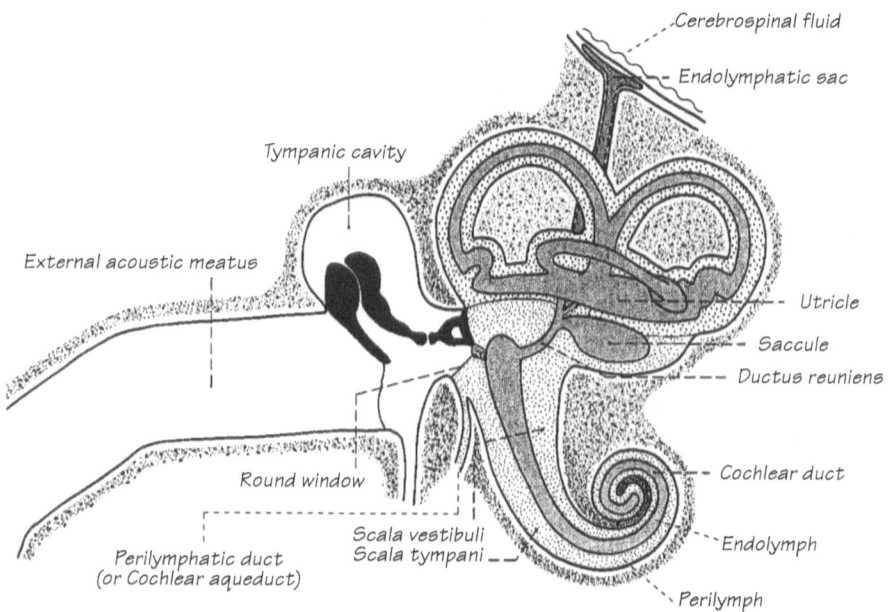

Fig. 7: The labyrinth.

Filled with watery lymph (endolymph), the inner ear is also surrounded on the outside by "perilymph," just as the brain and spinal cord float in cerebrospinal fluid. The perilymphatic space is directly connected with the cerebrospinal space via the perilymphatic duct or cochlear aqueduct; the endolymphatic space is in indirect contact with the cerebrospinal fluid via the endolymphatic duct (fig. 7).

After sound vibrations are received by the eardrum and relayed by the auditory ossicles, they appear in the lymph of the inner ear in the form of so-called *traveling waves*. According to Békésy, who received the Nobel Prize for his traveling wave theory, each sound triggers a traveling wave in the scala vestibuli. At the point on the cochlea where the wave "breaks" (that is, collapses in on itself and subsides into motionlessness), a sensory cell is stimulated. The length of the travelling wave depends on the frequency of the sound. High pitches produce waves that break at the beginning of the cochlea. The deeper the frequency, the longer the wave travels before breaking. Bass tones break in the

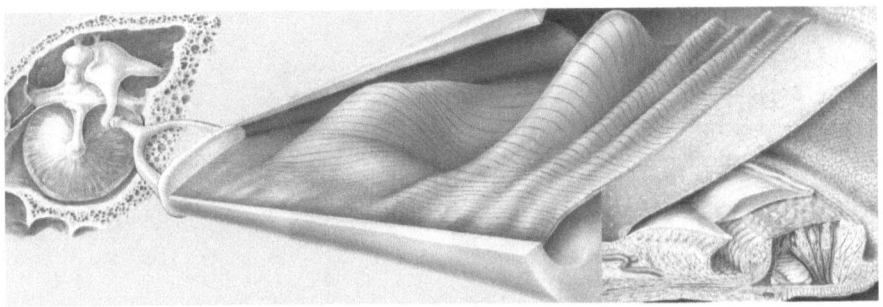

Fig. 8: Schematic representation of a traveling wave in the labyrinth.

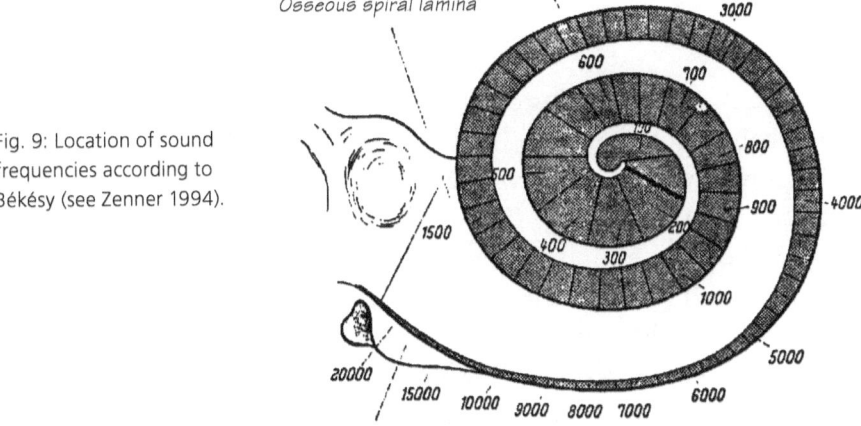

Fig. 9: Location of sound frequencies according to Békésy (see Zenner 1994).

apex of the cochlea. Under favorable conditions, as we may recall, a loud rock concert can still be heard at a distance of several kilometers—but only the bass! As in the cochlea, low and high pitches have longer and shorter ranges, respectively (figs. 8 and 9).

According to Békésy, the breaking waves of the inner ear lymph trip the fine hairs of the receptor cells in the organ of Corti, and the resulting depolarization of the sensory cells stimulates sensory nerves.[13] Two issues have remained unclear for a long time. First, the human ear is capable of distinguishing pitches with much greater accuracy than this mechanism permits. Second, three-quarters of the sensory cells

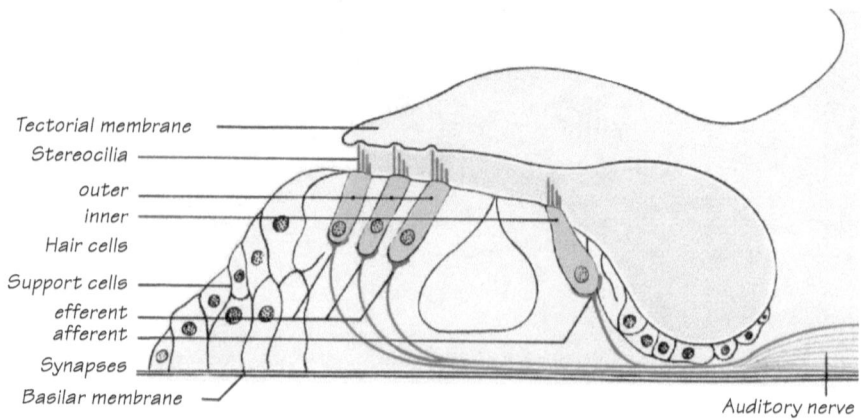

Fig. 10: Anatomy of the organ of Corti (schematic, according to Zenner 1994).

(the three *outer* rows of hair cells) are served primarily by *motor* neurons rather than sensory neurons. Only the hair cells in the innermost band, which is relatively isolated from the outer rows, are innervated primarily by sensory neurons (fig. 10).

Both these issues underwent further clarification when it was discovered that the outer hair cells have contractile cytoskeletons containing the muscle proteins actin and myosin in highly structured arrangements. For example, in the fine hairs (stereocilia) of the sensory cells, the actin filaments are almost *crystalline* in structure. The stereocilia are rigid, not flexible; and when they are displaced, the movement occurs around the joint-like connections where their severely narrowed bases are anchored in the cuticular plate of the sensory cells. Highly structured actin and myosin, however, are also found in the hair cell walls.[14]

When the passive traveling wave of perilymph breaks at the point in the cochlea that is characteristic of the frequency to be triggered, the sensory hairs of the outer hair cells are passively displaced. The effect of this displacement is to open pores in the sensory cells to the potassium-rich endolymph that surrounds them. Each cell, thus then depolarized, contracts as a whole through the action of the contractile protein prestin and leans (together with its embedded sensory hairs) in

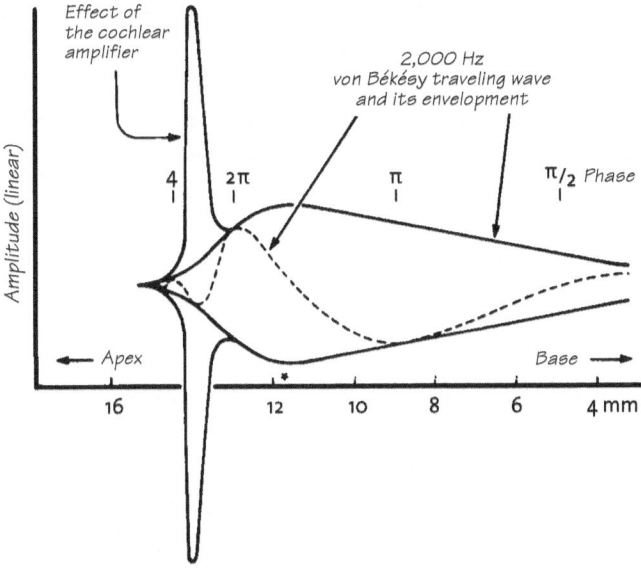

Fig. 11: von Békésy's traveling wave and its enveloping structure. The cochlear amplifier causes a high, sharp, frequency-dependent deflection of the basilar membrane (from Wever and Lawrence, as quoted in Zenner 1994).

the direction of the inner hair cells.[15] As a result, the tectorial membrane, which rests lightly on the tips of the sensory hairs, also moves. This movement in response to the travelling wave of sound is then transmitted by the endolymph to the sensory hairs of the inner, primarily sensory hair cells, which are then also displaced. In this case, however, the response is not muscular. Instead, the stereociliar displacement causes depolarization that is then transmitted further as an action potential in the neurons.

Sound reception in the inner ear, therefore, depends on subtle muscle activity. *The ear is a muscular sense organ.* The outer hair cells' active muscular response can be compared to the triggering of monosynaptic reflexes in muscles. For example, when a doctor's reflex hammer strikes the patellar tendon, the muscle spindle of the quadriceps is stretched, and the sensory receptors present there trigger contraction of the extended quadriceps via the spinal cord. The activity of the very fine muscles in the ear, however, is most comparable to the phenomenon of movement in the arms and legs. Through receptors in the muscle spindles and joints, we perceive a limb's change in angle and respond with muscle activity corresponding to the degree of extension—take for example, in reaction to the flexing effect of gravity when we are walking.

The active muscle response of the outer hair cells amplifies the traveling wave produced by perceived sound and intensifies the wave in its peak. As a result, the threshold of hearing is lowered by up to 40 decibels (dB), and the ability to distinguish between pitches is enhanced. This is why we speak of cochlear amplification and filtration in the inner ear (see fig. 11).[16]

(Muscle activity in the inner ear, as described above, is also presumed to be the source of so-called otoacoustic emissions—these are very subtle, very high-frequency sounds generated within the healthy inner ear.)

The quieter the sound, the more pronounced the amplification of the traveling wave. The *louder* the sound, the more the wave resembles the indistinct passive traveling wave that Békésy discovered in cadavers (see fig. 11). The changing sensitivity of the organ of Corti via motoric innervation of the outer hair cells plays a role in reacting to the loudness of sounds.[17] Here we see a physiological correlate of active listening.[18] Goethe captured this with astonishing precision in his study of acoustics: "To the ear, as an exalted organic entity, we must ascribe *reaction and call*, which make the mind independently capable of receiving and grasping what is brought to it from outside. And yet in the case of the ear we must consider conduction in particular, which definitely has stimulating and productive effects."[19]

From Technically Conceived Physiology to Living Physiology

Now let's consider the spatial scale at which these research findings are observed and calculated. Sensory hairs in the inner ear are displaced by only .01° or .4Å. "This is an extremely low geometrical value, *less than the dimensions of a hydrogen atom.*"[20] Can we still think of a movement smaller than the smallest atom in the periodic table as movement *in physical space*? Earlier in this chapter, we already mentioned that because of the placement of the auditory ossicles, their movement is not affected by gravity; and we proposed that the term "imponderability" or weightlessness is justified as a physiological property of sound vibrations. In addition, researchers have specifically emphasized that the ear's reaction (movement) is dependent on *the body's being alive*. The physicist Békésy discovered the traveling wave in figure 11 in cadavers, but active cochlear response (amplification) is dependent on the life of the organism; and it is very easily damaged by excessively loud noise or by certain medications (ototoxic antibiotics and loop diuretics).

These phenomena point beyond physical, earthly laws, indicating that the ear is integrated into the laws of the etheric element, or life element. In terms of physics, sound is a vibration in the air. That the ear's sensory receptors are not located on the eardrum, where they could be stimulated by *air* vibrations, but rather in the *fluid* of the inner ear, suggests that we do not perceive sound (via the ear) in the physical world at all, but rather in the living, etheric world. "The ear actually exists in order to overcome tones sounding in the air and to direct the purely etheric experience of sound back into the interior for us."[21]

In thinking that is adapted to life, we figure out the reality of the living organism by comparing phenomena and then grasping the "type" they have in common, as a living idea. Gisbert Husemann has done this for the phenomenon of the angle of displacement of sensory hairs in the inner ear.[22] The section that follows is based on his thoughts, which bridge the gap between physical and spiritual-scientific views of the hearing process.

Walking, Singing, Hearing

From the perspective of geometry, every limb movement consists in the rotation of a straight line (axis, shaft) around the joint's central point

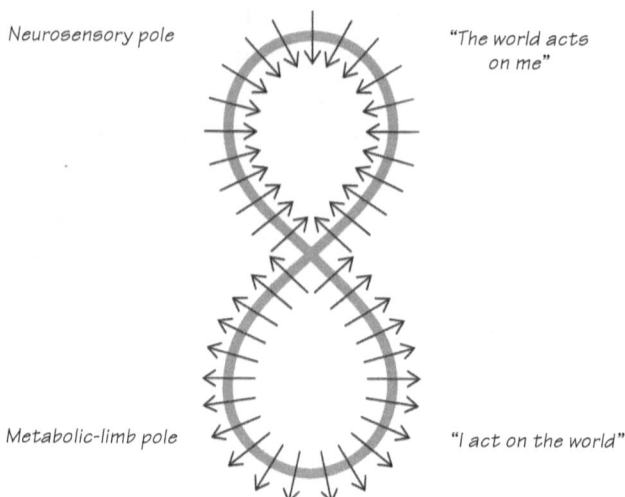

Fig. 12

by a specific number of degrees. This is why orthopedists use a medical goniometer to measure range of motion.

Illumined by the living idea of the ear's metabolic-limb character, passive and active displacements of the sensory hairs (stereocilia) of the inner ear are revealed as the smallest possible contraction and metamorphosis of limb activity. Arm and leg muscles move in interaction with gravity, whereas the inner ear's receptor muscles, like the auditory ossicles, move within the imponderable world of sounds.

The ear is a sensory organ and therefore *passive*; the limbs are *active*. The senses by which *the world acts on me* are concentrated in the head. In the limbs, *I act on the world* through what I do (fig. 12).

In considering the auditory ossicles of the middle ear, we saw that sound from the outside world sets these little limbs to moving in their joints. Here, sound substitutes for muscles. Only after this movement is transmitted to fluid does muscle activity set in, *inwardly-directed* and completely in the service of hearing.

Limb movement is movement in external space, movement that carries the physical body. When we hear music and speech, we experience

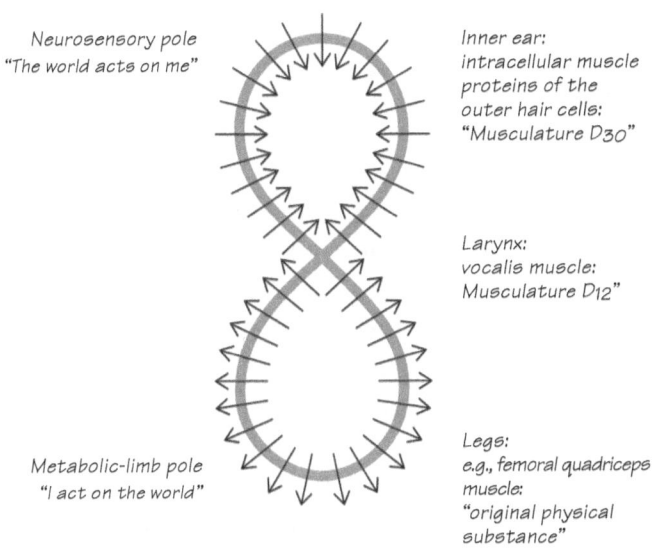

Fig. 13: The homeopathic potentization of muscle substance.

movement in *time*, movement that carries spiritual content.[23] Tones and sounds are carriers of soul movement and (in speech) of the spiritual movement of thoughts. The common factor in limb activity and inner ear activity is muscle movement. According to Rudolf Steiner, the polar metamorphosis of leg muscles into the auditory muscles of the inner ear is comparable to the transformation of substances through homeopathic potentization.

Because they are anchored in the earth's physical forces (gravity), the leg muscles are comparable to the original physical substance—that is, to the natural raw material used by pharmacists. In line with their spiritualized process, the muscles of the inner ear are "musculature D30," so to speak. They are "spirit-like" in Samuel Hahnemann's sense.[24] They work in the context of the sensory process but also *in the opposite direction*, from the outside in, again corresponding to the polar metamorphosis of homeopathy. A substance that is physically *toxic* in the lower body has *healing* effects when homeopathically potentized; that is, turned inside out in etheric counterspace.[25] In this sense, Rudolf Steiner calls the entire upper body the "homeopathically potentized lower body."[26]

A Symbolic Experiment

An experimental model may illustrate the metamorphosis of outer muscle movement into inner.

Imagine a rope with one end fastened to a wall. If we pump the other end, perhaps at approximately the speed of walking, we produce oscillations in the rope, waves that run up to the wall. If we continue to speed up the movement, we will ultimately reach our limit, and we will have to fasten the free end of the rope to a machine in order to continue increasing the frequency of vibration. We place the machine on a sliding bed so we can increase the tension on the rope as the frequency increases. At a certain point, without any change in the observer's position, the amplitude becomes so small that it becomes invisible to the eye. At that moment, however, when the rope's movement disappears from physical space, it appears to the ear: it begins to hum. As the speed continues to increase, the low tones of the subcontra octave (fourth octave below middle C) give way to higher pitches.

This symbolic experiment allows us to trace the path of movement (or will) from outwardly visible space across a threshold into the inner soul-space of musical movement and the word. *Extensive* movement becomes *intensive* movement.[27]

In the human body, this *moment of transition* between limb movement and ear activity also becomes an organic reality when striated, voluntary musculature retreats from the limbs' spatial world to appear internally, in greatly accelerated movements, as sound—in the *vocal cords* of the larynx. In the form of sound, this muscle movement still works *outward*, will-imbued, and to that extent it retains its limb orientation (see fig. 13). At the same time, however, it is also musical activity in the sensory sphere of the ear. Thus the sound-producing activity of the larynx muscles, supported by the breath, occupies the middle ground between the extensive spatial movements of the external limbs and the inner ear's muscular process of hearing.

Thus the location of the larynx corresponds to the crossing point in the lemniscates in figure 13. Having called the hearing process

"Musculature D30," figuratively speaking, we can now assign laryngeal activity to a middle potency suited to the respiratory system, such as D12.[28]

The Entire Muscular System "Hears"

Condon's high-speed camera studies of people filmed while speaking and listening to speech revealed that listeners, even newborns, produce subtle muscle movements, invisible to ordinary observation, in their whole bodies in synchrony with the movements of speech.[29] The listener's muscles "dance" to the sounds of speech! Because it provides actual experimental confirmation of Rudolf Steiner's research, anthroposophic researchers have taken up Condon's results.[30] This technique makes movements visible that, like inner ear movements, must be considered etheric rather than physical. These movements show that with their outer, physical, voluntary motor systems, listeners unconsciously make the micro-movements of the inner ear musculature on their own. The character of the inner ear's external hair cell motor system takes hold of the musculature of the entire body, and the body becomes "all ear."

Music and Mirror Neurons

The body's outer muscles "dance" to the movements of perceived speech, the inner ear responds to sound with muscle movements, and the metamorphosis continues *into the brain*. Seeing another person do something or speak activates not only the sensory (receptive) areas in my own brain but also the motor areas needed to carry out the perceived activity or speech; that is, the areas the other person is activating. As Molnar-Szakacs and Overy conclude, "Humans may comprehend all communicative signals, whether visual or auditory, linguistic or musical, in terms of their understanding of the motor action behind the signal."[31] The eversion through which imaginative thinking explains hearing is in complete harmony with modern physiology.

It has always been a puzzle that the labyrinth of the inner ear and the semicircular canals of the organ of balance are united in a single organ. In the context described above, however, it seems "bio-logical" in the literal sense that the sense of hearing, anchored as it is in the metabolic-limb system, should be allied with the sense for perceiving the body's spatial movements in the earth's gravitational field. In fact, the union of these two senses is proof that the polar metamorphosis of limb activity and ear activity is not simply "dreamed up."

Music Therapy

Against this background, the following insights into music therapy for movement disorders speak for themselves:

In a study of Parkinson's patients, sixteen patients received music therapy for three months while a control group of equal size received physical therapy. "Rigidity was more positively affected by physical therapy than by music therapy, but the reverse was true of bradykinesia (slow movement), which was significantly more affected by music therapy than by physical therapy. In contrast, neither therapy had any effect on tremor (trembling). Music therapy, however, also had significant positive effects on daily activities such as getting dressed, cutting food, frequency of falls, and so-called "freezing" (the temporary inability to move out of a fixed position), which is a common and disconcerting symptom of the disease."[32] Reportedly, for some Parkinson's patients who tend to "freeze" at traffic lights, the auditory cue provided by switching on a metronome in their pockets allows them to move promptly when the light turns green.

Parkinson-like movement disorders in post-encephalitic patients often also respond very well to music therapy, and even gait disturbances caused by alcohol intoxication can be eliminated through music. A neurologist observing guests at a party reported, "Although they were becoming increasingly drunk and *staggered whenever the music stopped*, their dancing seemed totally unaffected." In his book *Musicophilia: Tales of Music and the Brain*, Oliver Sacks gives this effect of music a chapter of its own: "Kinetic Melody: Parkinson's Disease and Music Therapy."[33]

The Eye Is Destructive; the Ear Is Motivating

When a patient with one arm paralyzed by a stroke practices movements such as making a fist with the paralyzed hand, comparing the movement to that of the healthy hand is supposed to provide both orientation and motivation. Usually, the sense of sight is used to compare the two limbs, but the comparison can also be conveyed to the patient through the sense of sound. Both arms are hooked up to an electromyograph, which registers the electrical potential generated by muscle cells. This electrical signal is converted into acoustic signals, which the patient hears through earphones. With eyes closed, the patient attempts to reproduce the "movement melody" of the healthy arm with the damaged one.[34]

This acoustic "bio-signal therapy" is far more effective than visual biofeedback. The therapist I interviewed at the Institute for Rehabilitation and Short-Term Therapy in Bremen said, "The eye is destructive; it overemphasizes the weak side's *deficits*. In contrast, the ear is *motivating*."[35]

Oliver Sacks describes his own experiences of rehabilitation after severe and nerve injuries to one leg: "And suddenly—into the silence, the silent twittering of motionless frozen images—came music, glorious music, Mendelssohn, *fortissimo*. Joy, life, intoxicating movement! And, as suddenly, without thinking, without intending whatever, I found myself walking, easily, joyfully, *with* the music. And as suddenly, in the moment that this inner music started, the Mendelssohn which had been summoned and hallucinated by my soul, and in the very moment that

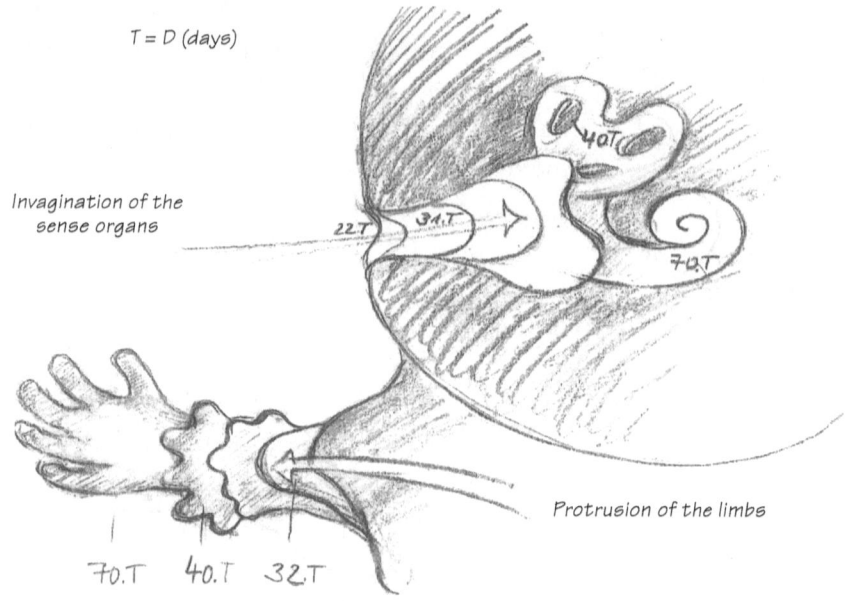

Fig. 14: Embryonic development of the hand and inner ear; D = days of gestation (drawing by Christian Breme [38]).

my "motor" music, my kinetic melody, my walking, came back—in the self-same moment the leg came back."[36]

The Proof of the Snake

The snake offers proof from nature that we hear with limbs. Snakes are reptiles, which generally have eardrums and middle ears. In snakes, however, the limbs have atrophied; not only the large limbs but also the eardrum, tympanic cavity, and Eustachian tubes, which have become deaf to air-borne sound.[37]

The overall picture of the motoric process of hearing, right down into the mirror neurons, as described above, sheds light on the Rudolf Steiner's findings: "In a very subtle way, the ear's sensory organization is inwardly associated with all of the nerves modern physiology calls 'motor nerves.' In reality, however, these nerves are also sensory nerves, and everything we experience as sound is perceived through nerve fibers embedded in our limb organization... That is why Schopenhauer and others have associated music so closely with the will."[39]

The Movement of Music from Note to Note

In the sector of our consciousness occupied by music, we not only observe its content dispassionately (as is possible with thinking) but also "go with" it inwardly, without actually having to move outwardly (as we would in dancing). The interpenetration of feeling and willing is evident in the fact that we can carry out movements inwardly, on the soul level, without doing so outwardly. We grasp rhythms, harmonic transitions, and the "gestures" of the intervals in a melody as willed movements, differentiating them inwardly on the basis of feeling. The archetypal phenomenon here is the "step" or interval from note to note.

In the major third interval from F to A, for example, we experience an inner musical quality that a number of musical thinkers characterize as a *gesture of movement or will*.[40] This will-aspect of music is what takes hold of people. It depends far less on their intellectual or educational background than is the case with literature or painting, for example. This is the world of the physiology of our metabolic-limb system, the system that produces energy. What we are calling "will" here must not be confused with its conceptual reflection, which we call "intention," "purpose," or "motivation." "Will" is the largely unconscious energetic layer of our being, at work when we accomplish actions. In everyday life, we experience it as an inner readiness to act. The organic basis of the urge to act is revealed by physical illness, which "dis-ables" us or makes us "unfit for work." To overcome illness (through fever, for example) the organism needs all of its metabolic and up-building forces. Once we recover, this body-directed metabolic energy is again free and the surplus is available for muscle movement, which is conveyed to the sense of life as readiness for action ("I feel well and fit again").

In patients with so-called endogenous depression, access to these independent forces is blocked for other reasons, so they experience a paralyzing lack of motivation on awakening.

The Reality of Music

Music allows us to experience and shape the movement impulses of this will layer, which we approach through feeling. We are now *inside* the realities we observed from the outside as we sought to clarify the metabolic-limb character of hearing. Schopenhauer put it like this: "Music is thus by no means like the other arts, the copy of the ideas, but the copy *of the will itself,* whose objectivity these ideas are. This is why the effects of music is so much more powerful and penetrating than that of the other arts, for they speak of shadows while music speaks of the thing itself.... In the lowest bass tones, I recognized the lowest steps in the objectification of will—inorganic nature, the massive planets. [Remember how creeping animals hear vibrations!] "And finally in *melody...I recognize the highest level of objectification of the will, human rational activity.... Thus melody tells the story of will illumined by deliberate consideration, of a will that leaves the imprints of its sequence of actions on reality...*"[41] (emphasis added). Rudolf Steiner confirms Schopenhauer's view of music (see pp. 20-21); in a different passage, he calls it an "instinctive Intuition."[42]

What is Schopenhauer suggesting with his "will illumined by deliberate consideration"? Day consciousness hides our will from us; in willing, we sleep. Only after death, when the human astral body and "I" are released from the physical and etheric bodies, do we awaken to our deeds of will, because their spiritual and moral essence is immortal.[43] A *spiritual sun* then illumines our will, the sun at work in Schopenhauer's words, "will illumined by deliberate consideration."[44] According to Schopenhauer, the spiritual sun of will-consciousness lives in melodies. This sun creates no images; it resounds. After death, in spiritual music deep within, we experience the movements, gestures, and intervals of our actions and steps. Did we apply our bodies' surplus forces of creation in harmony with their origin in the whole of creation, or did our actions work contrary to the forces that support and organize the world? That determines whether we experience consonances or dissonances in the "individualized harmony of the spheres" that we call our destiny (karma). Something of the spiritual cosmic music of life after death resonates in the music of great composers who are able to penetrate such depths of inwardness. This is what prompted the pianist Alfred Brendel to confess, "In parts of [Mozart's] *Mass in C Minor* and the choruses of Idomeneo, in the commander's "Maestoso"; or in the *Adagio and Fugue in C Minor* (KV 546), Mozart's music no longer plays a role. These pieces confront us like destiny itself, exalted and relentless; no best friend and comforter, no bringer of the blissfulness of death, but the other—the superior power that strikes us dumb. What is achieved here, not only in formal perfection, but also in power of feeling, is beyond human."[45]

Through such music, Alfred Brendel felt transported to the threshold of the spiritual world. In regard to the other side of that threshold, Rudolf Steiner stated it like this: "It is true that the moral aspect of any action performed on your path, whether of good deeds or bad, is transformed into sounds after passage through the portal of death; not immediately, but after a certain time. You then *hear* your moral behavior here in the earthly world. Your morality becomes beautiful music, your immorality ugly music. And in these harmonious or dissonant sounds you hear words, as if spoken about your actions by the higher hierarchies sitting in judgment."[46]

Karmic Development and Eversion

After death, experiencing the spiritual will-structure of our actions in the form of music is associated with the process of *eversion*, through which the forces that shape destiny sow the spiritual seeds of the physical, living body for the next earthly life. How do we experience our deeds after death? What we have done in the physical world comes to meet us from outside, in a process of spiritual sensing. We *endure* our deeds or *experience* their positive effects: deeds of will become spiritual sensory experiences.

In music, we experience the process that was described earlier in functional and sculptural terms as the eversion process through which limb movement metamorphoses into the movement of hearing (see fig. 13, p. 15).

This eversion is an image both of what we experience in the spiritual world through music during the time between death and rebirth and of what we create in the course of experience together with the beings of this spiritual world. Through a process of eversion, we transform the actions of our limbs into the spiritual primordium of our head for the next life. According to anatomist Jochen Staubesand, the skeleton offers surprising evidence of this fact.

The Everted Will in Hearing and the Ear

Staubesand pointed out that in the human skeleton, the shape of the skull is highly individualized, often with striking discrepancies between it and the rest of the skeleton." The thickness of the skullcap is subject to significant individual variations. The thickest skull walls can be nearly three times as thick as the thinnest, *without the rest of the skeleton displaying any exceptional robustness*. Mechanical functional features fail to explain this striking variability."[47]

In mammals, there is no evidence of similar variability and especially no comparable discrepancies between the skull and the rest of the skeleton. In animals, the structure of the skull corresponds to that of the rest of the skeleton; a single consistent formative principle is at work in limbs, torso, and skull.[48] Rudolf Steiner pointed to this phenomenon as early as 1910. "In the human body, we are most interested in the skeleton's noblest organ, the skull; and especially in its sculptural qualities. Skull structure is different in each individual because it remains open to the underlying individual element of the 'I.' Animal skulls, however, express the characteristics of the species."[49]

This variability in skull structure, evident only in human beings, is

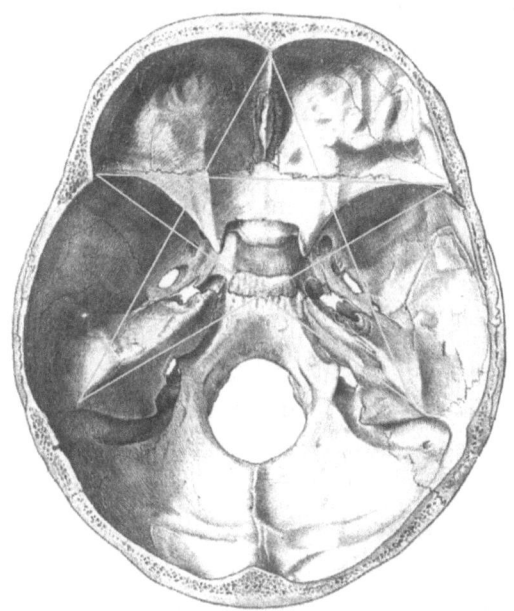

Fig. 15: The base of the skull (according to Toldt/Hochstetter).

brought about by the process of inversion that takes place between death and rebirth. The individual animal, lacking an inherent I-organization or individualizing principle, is not subject to this process. In humans, however, the I-organization transforms the spiritual formative forces of destiny into the shape of the new body. Spirit research describes in concrete detail how the legs and arms, in the nexus of spiritual forces derived from their actions, are "turned inside out" and reincarnated as the *auditory ossicles* in the next life. The foot and calf become the hammer, the kneecap the anvil, and the thigh the stirrup.[50]

Hearing is active in an invaginated will process, and the reality of music is the earthly reverberation of the forces that shape destiny. We can grasp this statement by considering the base of the skull as an image (figure 15). We see there the pentagram of the cosmic human being, with the organs of hearing and balance incorporated into the "legs" of the petrous bones. The "arms" extend outward through the light of the eye sockets, and the most physical sense (the sense of smell) forms the pentagram's "head." Because the skull is derived entirely from cosmic formative forces and completely overcomes the effects of physical forces, "imaginative forms condense to the point of becoming physical in the shapes of the head."[51]

The Inner Ear and the Intestines

In the ear, the stirrup rests against the cochlea of the inner ear. Similarly, the head of the femur abuts the fluid-filled loops of the intestines in the abdominal cavity. In cattle, so Rudolf Steiner tells us, the intestines are archetypally developed and reflect cosmic forces with exceptional purity[52] (see fig. 16).

This archetypal image of an intestine allows us to understand why Steiner indicated that the fluid-filled human cochlea is the re-embodied intestine.[53] We began with the intimate connection between hearing and the metabolic process of chewing and swallowing, and now we return to it. Each section of the gastrointestinal tract has different ways of analyzing the chemistry of substances and breaking down and absorbing them. Similarly, the different pitches we hear are perceived in different sections of the cochlea.[54]

"If inherent possibilities were all that came into play, the primordia of the arms and legs of an embryo in the womb could develop into ears. These primordia definitely have the potential to become ears, which means that it would also be possible for the human body to develop not only with one ear here and another here [It is clear that the speaker is pointing to his own ears.], but also with one down here. This sounds paradoxical, but it is full of truth. It would be quite possible for a human body to have an ear down here. So why doesn't an ear develop here? Basically, because at a certain point in its embryonic development, it enters the realm of earthly gravity.[55]... The force of gravity pulls on what wants to become an ear and reshapes it. Under the influence of earthly gravity, the ear that wants to develop down below becomes the lower part of the human body."[56]

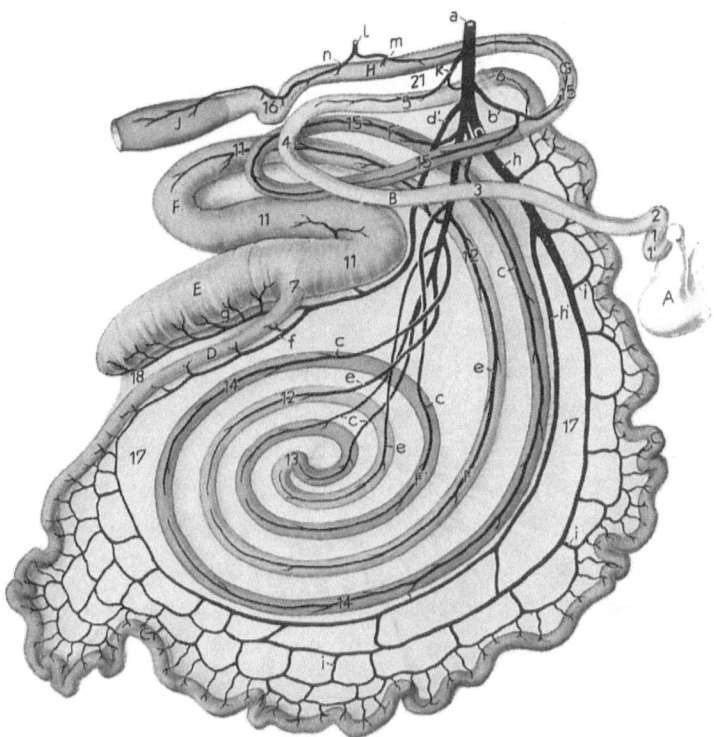

Fig. 16: Cattle intestines.

Based on the metabolic-limb nature of music, Rudolf Steiner recommended that teachers who sing with children imagine themselves (artistically) transported into the digestive atmosphere of a herd of cows! "A herd of cows that have eaten and are lying there in the pasture; the digestive process of a herd is really something quite wonderful. In the cow, something like an image of the entire cosmos is present." When singing with children, teachers should feel this sensation of wellbeing in the inner sequence of notes. "That is how you must allow the child to experience these little moments of bliss. You must really evoke a feeling for music in your whole body and enjoy it thoroughly yourself."[57] This is how teachers infuse their feeling for music with cosmic metabolic wisdom, which is the basis of *will*.

The musical origin of the digestive system

Let us look all the way back to the physical body's initial state as spiritual science describes it during the earth's first stage of evolution, the so-called Old Saturn state.[58] Of primordial physical existence, which must be imagined as still etheric and consisting of heat, Steiner wrote: "It may seem sheer folly to many when we speak of what then reveals itself to suprasensory cognition. Saturn's interior was like surging, intermingling sensations of taste. Sweet, bitter, sour, and so on could be observed at various points within Saturn, while outwardly all of this reverberated into celestial space as sound, and was perceived as music of a sort."[59] Nature itself, however, occasionally reveals what may seem "sheer folly": A flutist in Zurich had legitimate synesthetic perceptions of taste specific to each of the intervals of the diatonic scale. At the University of Zurich's Institute for Neurophysiology, these simultaneous perceptions were investigated using every trick in the book. Here are the "qualities of taste" and their triggering intervals:"[60]

Interval	*Sensation of taste*
minor second	Sour
major second	Bitter
minor third	Salty
major third	Sweet
Fourth	Grassy
augmented fourth/diminished fifth	Disgusting
Fifth	pure water
minor sixth	Creamy
major sixth	Milky
minor seventh	Bitter
major seventh	Sour
Octave	no taste

The chorda tympani, the nerve that serves the taste buds of the front of the tongue, also runs through the middle ear—an outer link between hearing and tasting. Gisbert Husemann called this phenomenon "a bookmark in nature's book."[61]

CHAPTER 2

Music Is "Chemistry from the Inside"

The Fluorine Process in the Human Being

> *The organism's essence lies in its action, not in its substances. Its organization is an activity, not a construction of substances.*
>
> ITA WEGMAN / RUDOLF STEINER[62]

Upon opening a chemistry textbook, the first thing we read about is the periodic table of the elements and electron pair theory. Elements strive to achieve a stable configuration in their outermost electron shell (valence shell). In period 2, for example, oxygen (with six electrons in its valence shell) prefers to combine with partners that contribute two electrons, giving the oxygen a total of eight electrons and the same configuration as the inert gas neon. For example, two hydrogen atoms complete the configuration, giving oxygen eight electrons in its outer shell: H_2O. To achieve the configuration of neon, nitrogen (with five electrons) needs three hydrogen atoms, yielding NH_3; carbon needs four, forming CH_4.

The attempt to achieve the same number of electrons as the next *higher* inert gas, however, applies only to substances in group 4 and up. Elements in groups 1, 2, and 3 (which have only one, two, or three electrons in their valence shell) have the opposite tendency; they combine with partners who take away their extra electrons so they themselves are left with the same configuration as the next *lower* inert gas. Lithium, beryllium, and boron aim to achieve the same configuration as helium. They are electron donors or reducing agents and base-forming elements. Their opposites are electron acceptors, oxidation agents, and acid-formers. This dichotomy sheds light on the unique position of carbon, located in the *middle* between these two tendencies: carbon is capable of both

Music Is "Chemistry from the Inside" 31

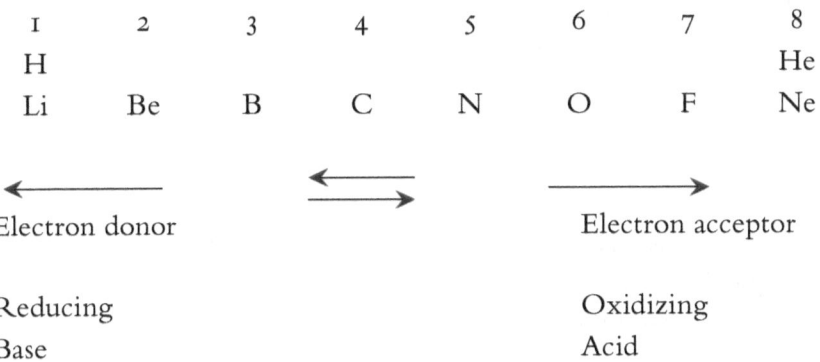

Fig. 17

gestures. Its universal importance in the organic chemistry of living organisms is based on the fact that it can enter into combinations in either direction (helium or neon, fig. 17).[63]

When we actively pursue this train of thought, feeling is also engaged and music awakens in our thinking. Musical thinking then recognizes that the relationships among elements described here are also evident in the seven intervals of the major diatonic scale. The second, third, and fourth relate only to the *tonic below them*. Beginning with the fifth, the remaining intervals also relate to the octave above. This tendency becomes more noticeable in the major sixth and peaks in the major seventh. The discoverers of the periodic table were already aware of this analogy to music.

In 1862, Béguyer de Chancourtois used the combining weights of elements as the basis for arranging them in a spiral wrapped around a cylinder. Elements with similar properties lay one above the other. In 1865, Newlands formulated the chemical "law of octaves": When the elements are arranged according to increasing atomic weight, similar physical and chemical properties recur after each interval of seven elements, as in musical octaves.[64] The noble gases, which were discovered only after 1895, rarely form compounds with other elements because their valence shells are full, so they cannot contribute electrons to or accept them from partner elements.

There is one difference between chemical and musical octaves that could not yet be understood in Newlands' time: a musical octave is also always the starting point of a new scale, that is, the octave is also the tonic. Something of the essence of the mineral state of matter appears to emerge in that the noble gases, as octaves, fall out of the course of events. That is why the "helium state" that lithium attempts to achieve by reacting with its partners is inherent in lithium itself, just as the "neon state" is inherent in fluorine. The "octave" process means that elements transcend themselves in combination with other elements.

Musicians are familiar with the energy levels that intensify incrementally and discontinuously from the fifth to the sixth and finally to the seventh step of the scale; they are reflected in the finale of every diatonic composition. The moving and resolving energies of the finale of a sonata or symphony are nothing other than the forces of the seventh being played out until they resolve in the octave of the final chord, which is like an inner experience of the stable, non-reactive state of the noble gas configuration. A proven method of holding out against the finale's forces of resolution is the fugal style which Beethoven uses, for example, in the final movement of his Piano Sonata, Opus 28. He maintains the fugue through four entries; the fifth requires octave doubling in the left hand in order to prevail against the resolution. The process dissolves into sixteenth-note arpeggios in the sixth step; in the seventh, the figure is shortened to repetitions of the sixteenth-note figure of its closing passage; in the eighth, the movement reaches its goal in a compositional elaboration of the octave:

Phases

I

Example 1: Ludwig van Beethoven, Piano Sonata, Opus 28, fourth movement, measures 79 -113.

Music Is "Chemistry from the Inside" 33

Phases

Phases

When a symphonic finale is in the stage of the seventh interval, the increasing tempo and the "oxidative" resolving tendency of the dominant seventh chord require an appropriate counter-response; the physical aspect of the music is intensified. The volume increases in a crescendo to *forte* and *fortissimo*. Large instruments that did not appear in the first three movements are now added—more string basses, bass tubas, trombones. The energy of the music encroaches on the world of physical forces; the musicians begin to sweat as their work becomes increasingly physical. These are examples of the physical effects of the inner musical experience that is embodied in fluorine, as an example of the seventh group, in chemistry. The other halogens, chlorine, bromine, and iodine, should also be examined from this perspective. In homeopathy, the drug pictures of the halogens show many similarities.[65] Both iodine and the thyroid function it serves (which this author describes in a different book) are related to the oxidative physiology which we experience from inward in the musical seventh.[66]

The Fluorine Process from the Chemical Perspective

"Chemically, fluorine is the strongest oxidation agent, the one with the strongest chemical activity. That is why fluorine reacts with all the elements except gold and platinum. In fluorine, even water burns with a bright flame."[67] Fluorite, CaF_2, owes its German name, *Flussspat* ("flux spar"), to the fact that it lowers the melting point of slag and is therefore used as a flux in smelting ores. Hydrofluoric acid is such a strong solvent that it even dissolves glass, so it cannot be stored in ordinary test tubes.

The analogy between the periodic system of the elements and the arrangement of musical perceptions in the seven-note scale also deserves to be taken seriously because it immediately became apparent when the periodic system was discovered, although from a different perspective. Rationally relating chemical and musical processes, however, poses a methodological challenge. It is difficult to see any relationship between the contexts for these two areas of our experience; chemical substances on the one hand, and the inner activity of musical experience on the other. The analogy makes sense only in the context of the metamorphosis of etheric formative forces, which we will explore in chapter 3. Rudolf Steiner endorsed Newlands' discovery, explaining that "the image of the octave can help us visualize the arrangement of elements in the periodic system. We see an analogy between the inner laws of pitch and the structure of matter as it prepares to be active in chemical processes. Thus we are also justified in seeing the processes of union and dissolution in material existence as an outer image of an inner cosmic music, which reveals itself to us as earthly music only in one particular instance."[68]

In the previous chapter, we understood earthly music as an etheric event in connection with metabolic processes; that is, as the chemistry of the intestines metamorphosed into the ear. As such, it reflects the laws of the music of the spheres. It is a living afterimage of spiritually formative music, as we saw from the example of Beethoven's Sonata, Opus 28.

The Seven Stages of Life

The fact that the weight of bodies on Earth causes them to be pulled in the direction of the Earth's center is not a superstition. Similarly, it is no mere superstition that all development or evolution in the living world takes place in seven stages. The sevenfoldness of the etheric world is reflected in many physiological phenomena.[69] Here, this fact is of particular interest in regard to the chemical ether, which governs all chemical processes. "Thus we have seven colors in the rainbow, just as there are seven notes in the scale and seven levels of atomic weights in the atomic realm."[70]

"The mathematical proportions of chemistry are really expressions of the mathematical proportions of the music of the spheres, which has become mute by condensing into matter."[71] To make this connection fruitful for a musical approach to chemistry, we must first take a look at the evolutionary history of matter.

The Genesis of the Ethers and the States of Matter

Rudolf Steiner presented a spiritual-scientific cosmology in his book *An Outline of Esoteric Science*.[72] From it, we learn that at the beginning of its evolution, planet earth first manifested physically in pure states of heat.[73]

This "Old Saturn stage" of the earth was followed by a second stage, the "Sun" stage. As the "Old Sun," the earth manifested physically on the gaseous level and was also enlivened. After the Old Sun had spiritualized at the end of its evolution, the earth emerged transformed from the spiritual world of creation in its third embodiment, "Old Moon," which reveals to spiritual research not only physical and living processes but also the feelings, emotions, and soul life of its inhabitants. At this stage, the physical element had condensed to the fluid level. At the end of its time, the Moon stage of the earth's evolution also died away, and the earth entered its fourth incarnation, this time as the mineralized planet we know today. In addition to living and ensouled organisms (plants and animals), this stage also shelters human beings, who are spiritually active and self-aware.[74]

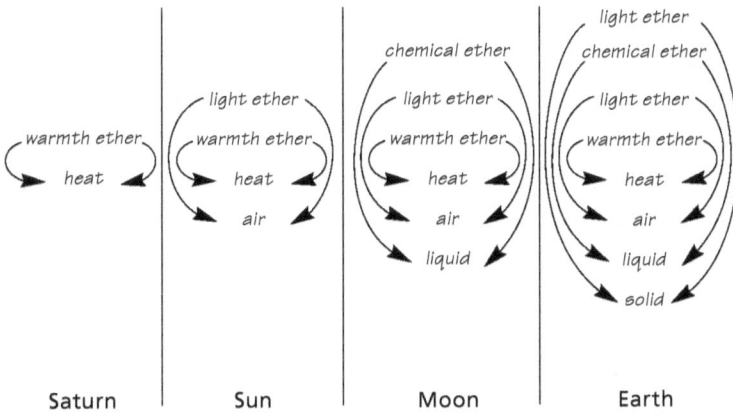

Fig. 18

The incremental condensation of physical substances out of their initial state of warmth into the gaseous, fluid, and solid states required a corresponding enhancement of the etheric forces of life and a transformation of their activity. "Heat" in the physical sense and "warmth ether" are still like two aspects of the same thing. The airy organisms of the Old Sun were enlivened by the light ether, which we must understand as a further development of the warmth ether.

The etheric forces of life underwent the next step in their evolution during planet Earth's third incarnation, Old Moon. This development occurred when exalted cosmic spirit-beings sacrificed an aspect of their soul-body and incorporated it into the living human organism, thus endowing the ancestors of humans with the capacity for sensation and for experiences of pleasure and displeasure. These beings are called *dynamis* in ancient Christian esotericism and "spirits of movement" in the modern terminology Rudolf Steiner selected on the basis of his own spiritual research. The spirits of movement took hold of the light ether from the inside, transforming it and raising it to the level of the chemical ether, which allows fluid substances to react with each other, and which is active in all combining and dissolving of substances. The chemical ether, therefore, is like a child that emerges from the union of the *dynamis'* astral body and the etheric human element. That is why music and chemistry are two manifestations of the same astral forces. There is only one nature.

When we say there is "bad chemistry" between two people, our language is pointing to this connection between human soul life and the chemical ether. Earlier chemists spoke of "elective affinities," which is why Goethe, in his novel of the same name, applied this term to a constellation of relationship and destiny and its transformations. In a self-published advertisement for *Elective Affinities* (1809), he wrote, "It seems that the author was led to this strange title by his advanced studies of physics. He must have noticed that in teaching about nature, ethical similes are often used to bring something that is far removed from the sphere of human knowledge closer to hand, and so in this moral instance he also must have hoped to *trace a chemical figure of speech back to its spiritual origins*, and all the more so, since everywhere there is only one Nature; and even the realm of happy freedom from rationality is inexorably crisscrossed by the tracks of dreary, passionate necessity, which can only be completely extinguished by a more exalted hand, and perhaps not in this life."(emphasis added)[75]

The Number Ether

The organic function of the chemical ether has a further consequence for the phenomenon of the physical body: The chemical ether structures the physical body according to number. "The astral body 'counts' the etheric body, shaping it through counting."[76] A third, inner manifestation of the chemical ether as it appears to the sentient soul is sound. Rudolf Steiner used all three terms (number ether, sound ether, and chemical ether) more or less interchangeably. In the sound ether, the chemical ether, which is otherwise at home in the fluid element, manifests through the air (see pp. 82 ff.). Figure 19 illustrates the relationship among all three manifestations.

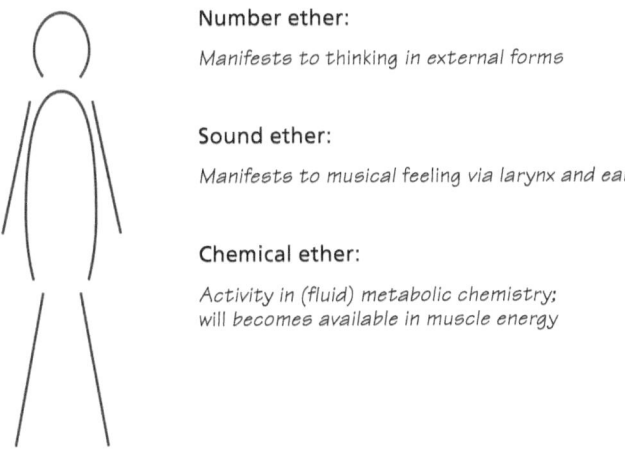

Fig. 19: The chemical ether and its metamorphoses.

The Periodic Table as a "Creation Document"

The seven notes of the scale, which are reflected in the seven groups of the periodic table, reflect a universal etheric law. All evolutionary processes take place in seven stages, which is why E. Bindel and A. Blickle call the periodic table of the elements a "creation document."[77] Our modern experience of the seven intervals reflects the fact that as human beings we have already undergone four of the seven stages of Earth evolution. They lie in the evolutionary past, which means they have *become our bodies*. We live in physical bodies as a result of the "cosmic prime" of the Old Saturn state, and in life-bodies because we have passed through the Old Sun stage. Ever since the Old Moon stage with its world-shaping forces of sympathy and antipathy, what had been the mere stream and flow of life (in the interval of the second) has been opened up to inner feeling (in the major and minor third). We are now living in the specific effects of the fourth.[78] We owe our "I am" consciousness to this fourth stage of cosmic evolution. The intervals of the fifth, sixth, seventh, and octave point to future stages of human evolution. Although they are present as forces in the cosmos, human beings have not yet taken possession of them in the same way.

The Melody of Life and the Actions in the Outer World

How do we experience the musical quality of the fifth from within? In the first four steps, we shaped our own being and finally secured the "I." In the fifth, however, we open our eyes and awaken to the world, touched by something that streams toward us like a breath of air touching our skin. Thus Rudolf Steiner's characterization of the fifth includes the comment that it is the interval of the skin—that is, of sensory surfaces.[79]

Sensing the gesture of the major sixth is like moving outward toward something in the world to which we aspire with longing. In many countries, therefore, the sixth is considered the interval of love. In major and minor sixths, our connection to what we perceive in the world is tinged with pleasure or pain, respectively. In the seventh, we are set in motion, breaking down the boundary between ourselves and the world—we actively take hold of the world and *transform it*. The octave then contains the feeling of accomplished fact. Rudolf Steiner's first early suggestion for the octave gesture in eurythmy was to touch something in the outer world with the palm of the hand and then turn the hand around.[80]

What we have become now lives in us as the fourfoldness of the first four intervals: prime, second, third, and fourth. The fifth, sixth, seventh, and octave are our future, our will-and-limb organization in its connection to the world. Here is one example: we see a flower. If we simply pass it by because at the moment our soul is incapable of developing a connection to it, the flower remains on the soul level of the fifth—that is, a perception. But if we stop and admire the flower's beauty, the mood of the sixth arises in our soul, and if as a result of this devotion we pick the flower and give it to a person we love, our soul actively takes hold of the outer world in a "seventh" process as we move into action. Finally, the connection to the world that was awakened in perception comes to rest in the octave.

Another example: I am sitting on the streetcar looking out the window when I spot someone I know out on the street (fifth). It occurs to me that I still haven't sent him an address I promised to send quite some time ago. This is an inner perception (fifth) that takes place in memory and thinking. Together these two perceptions arouse a feeling of embarrassment and discontent or shame (minor sixth). Back at home, this emotional process passes over into the seventh; that is, I immediately write a letter of apology with the address I promised to send, write the address on the envelope, and put the stamp on. I leave the house again and go to the mailbox. As soon as the letter falls through the slot, the octave "resounds" in my soul. The process that began with seeing my acquaintance has been resolved as far as I am concerned. For me, it has been put to rest, but in the world my action continues to have effects. In the octave, my action (that is, some aspect of myself) has become a world process.

Actors must master this scale with precision because their power to convince depends on it. The "hero" stands alone on the stage. In an inside jacket pocket, he holds a dagger firmly in his fist. He has resolved to kill his adversary. As he stands alone on the stage, brooding, reliving all the disgrace he has experienced, he is in the stage of the tonic to the fourth—a self-involved human being. But now a lady, his beloved, enters. The audience waits; when will he awaken to the fifth and notice her presence? In the dialog that follows, it is important for the audience to know that this "beloved" is a fake; she is really his adversary in disguise, full of evil intent. The audience wonders when the hero will recognize that he is mistaken about the connection he feels, that he is the victim of a false major sixth. When our hero sees through the disguise, his real relationship to the other flares up (minor sixth). A brawl (seventh) ensues and the turmoil is resolved in a horrific murder (octave).

Here we come full circle to previous chapters, where we discussed the musical character of actions (see p. 22). "The human being, when making the transition to activity, is actually structured like a musical scale."[81] Thus tone eurythmy which transforms inwardly experienced movements of music into outwardly visible, artistic limb movements, is an art derived directly from the nature of the human being. In eurythmy's gesture-images of the intervals, however, what we characterized as a willed relationship to the world is held back, making the connection to the periodic table's levels of energy and combination potentials all the more apparent. The eurythmic major sixth is imbued with the breath of oxygen, the gesture of the seventh with the fragmenting energy of the fluorine process, which comes to rest in the octave by finding a partner in the salt process. In the periodic table, the "tetrachord" of the fifth, sixth, and seventh drives oxidative processes that release the energy and warmth of movements of will (see p. 32).

A Second Human Being Emerges

Let's trace this threefold will-relationship to the world back to the physical body. The fifth (perception in sensory activity or in one's own thinking) is embodied in neurosensory processes. In the major or minor sixth, we open ourselves to the world on the feeling level through the middle realm of the breath and the heart. In the seventh, movement is transferred to the outer, physical world via the metabolic/limb system, with which we accomplish actions.

In our relationship to the world, as we see, a second human being emerges, taking shape continuously out of our feeling and doing.[82]

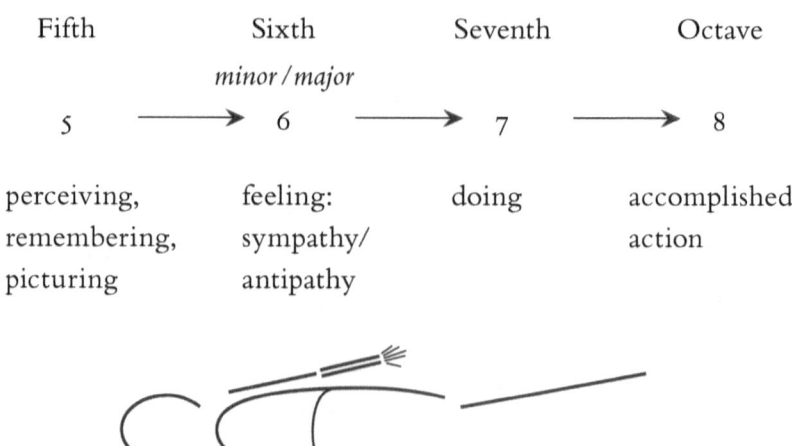

Fig. 20

Music Is "Chemistry from the Inside" 43

Fig. 21: Jacob's ladder, from the Wenzel Bible (Codices Vindobonenses 2759-2764, folio 27r).

Fig. 22: Rudolf Steiner's Faust figure from the ceiling fresco in the small cupola of the first Goetheanum. Between body (skeleton) and spirit (angel), receiving his higher "I" as child from the future.

If we look at this emerging being with the vision we applied in the previous chapter to the musical process of the development of destiny, it becomes apparent that the step from the seventh to the octave is a musical image of crossing the threshold into action. In other words, in the sensation of the octave lies a spiritual potential, a germ of the consequence of destiny or karma that is created as this action is accomplished. The effects of my actions persist in the world even after I die, and I will encounter them again in my next earthly life. Rudolf Steiner predicted that the sensation of the octave would continue to evolve and intensify along with our evolving capacity for grasping our own destiny.[83]

Destiny is the will of our higher "I." Together with the hierarchies, we shape this destiny in the divine-spiritual world during life before birth. Sensing the octave has also always included sensing the higher "I," the higher human being who lives in divine will and with whom we aspire to act in concord. In the choral words for his *Ninth Symphony*, Beethoven wrote:

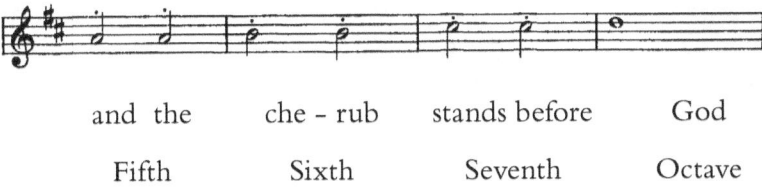

| and the | che - rub | stands before | God |
| Fifth | Sixth | Seventh | Octave |

Example 2

Figure 21 (p. 43) is an illustration of Jacob's ladder from the Wenzel Bible. The seven steps of the astral body rise above Jacob's head into the cosmos as he lies asleep and dreaming. The *developed* being of the lower tetrachord wears the colors of Jacob's physical body. The *developing* being comes toward the earthly "I" from the future, sent by the higher "I" of God. This old musical image becomes a spoken image in the figure of Faust as Rudolf Steiner painted it in the small cupola of the first Goetheanum (fig. 22).[84]

In this sense, our destiny approaches us from the world as our fate, which is also our own actions. On the physical plane, the world rejects actions that it cannot incorporate in meaningful or useful ways. Defective cars are recalled to prevent accidents. As a first-grader in a Waldorf School, the author attempted to knit a potholder, but his teacher unraveled it because it was "stiff as a board." Because it could not be wrapped around a pot handle, it was not allowed to cross the threshold into the outer world of use.

Finger Dexterity and Dental Health

"When you visit a craft lesson in our Waldorf School, you find the boys learning to knit alongside the girls, and boys and girls all doing the same things. We do this because their fingers need to become skilled and flexible; their souls need to be driven into their fingers. By driving the soul into the fingers, we encourage everything related to dental development. It makes a difference whether we encourage dexterity or allow children to grow up unskilled. Any neglect in this area will show up later in premature tooth decay, although of course more so in some than in others. On the macroscopic level, the actions and interaction of the hands and feet correspond to the effects of fluorine. The fluorine effect is not what we imagine it to be from the atomistic perspective. Rather, it is the constitution that develops when the fingers acquire dexterity and mobility."[85]

The eurythmy gesture of the seventh is a musical image of ensouled finger dexterity, an imaging technique for the fluorine process in the chemical ether. But now the questions arise, why and how does this seventh-related fluorine process strengthen the teeth? The development of dental enamel and bone in the body, which drives organic substance to the very border of lifelessness, is one pole of life, while the necessary counterpole involves the dissolution and regeneration of form. These two processes are present everywhere and at the same time, but are not always in balance. For example, the "form pole" dominates in the head, where living, etheric, and soul-spiritual human forces are released from organic activity; that is, from the rigidifying substance of the head. This is the same process that takes place in the entire body at death. In the teeth, fluorine combines with calcium phosphate to form apatite, the stable octave or final chord. Similarly, table salt forms the "octave" of the chlorine process.

Since in the process described above, the body as a whole does not die, we owe our waking consciousness to this "death in life," which is the mystery of our neurosensory activity (see pp. 68 ff.). In the opposite process, life's formative forces take hold of physical substance in the metabolism and limbs, enlivening that substance and repeatedly rebuilding the body. Older physiological texts spoke of the "principle of salt" of the upper body and the "principle of "sulfur" of the lower. Every substance, including fluorine, oscillates between these poles in a mercurial process. Fluorine's dissolving, sulfuric aspect appears in the lower body, right down into the fingertips, where it appears in the form of limb mobility enhanced to the level of the seventh. In the head pole, which shares its mineralizing tendency with the entire skeleton, the fluorine process has a saline effect, functioning as the salt (fluorapatite) that gives calcium phosphate its greatest degree of acid resistance.

The mercurial instability of the fluorine process is evident in the increased incidence of dental caries during pregnancy. An old proverb states that "mothers lose one tooth for each child," and a recent study has confirmed this fact.[86] Within the mother's abdomen, the child's new body develops (that is, condenses to the point of ossification). Relative to the "sulfuric biotope" of the maternal abdomen, this is a salt process, and it is also a "foreign body" within the maternal body. This is why the mother's etheric body loosens and disengages somewhat from her physical body: "The result of the freeing of the etheric body that takes place at this end of the human organism is to pull the etheric body into the organism at the other end, where it is also associated the opposite effect, namely, destruction."[87]

Learning to knit and crochet drives the soul into the fingers; balancing and writing with the feet drives the soul into the toes, and ties up the fluorine forces in sulfurous activity in the lower body. In contrast, in children who are clumsy and sluggish, these same fluorine forces tend to stray into the upper body; specifically, into the teeth, where their displaced sulfurous, inflammatory activity produces dental caries.

We owe the polar structure and functioning of the human body to our I-organization, which inserts the metabolic-limb processes of one earthly life (to the extent that they persist as spiritual processes after death) into the head organization of the following earthly life. Thus all of the organs below the diaphragm have their counterparts in organs within the head organization. To spiritual researchers, therefore, teeth are re-embodied fingers and toes.[88] The forces of the previous life's toes reappear in the tooth-forming forces of the lower jaw, the forces of the fingers in the teeth of the upper jaw. As mentioned on page 25 above, the hearing process of the hands and feet are transformed into the hammer of the ear in the next earthly life. Finger movements, with their activity directed out into the world, become the teeth.

If we dare to imagine a physical disorder such as dental caries interacting with spiritual realities in this way, we look for proof in the realm of physical facts. In one study, physical examinations of 373 adolescents aged 13 to 16 years included checking for dental caries.[89] Standardized DMFT (decayed, missing, or filled teeth) values were used to determine that children attending an anthroposophic school for the handicapped had significantly healthier teeth than children in state schools, although neither group had received prophylactic fluoride treatments. Or had they—in the sense of the "functional fluorine process" as Rudolf Steiner understood it?[90]

Fluoricum Acidum for Severe Insomnia Due to Restless Leg Syndrome

A musical understanding of the fluorine process also has therapeutic applications, as the following case study demonstrates. During our third trimester at the Eugen Kolisko Academy, we were introduced to a 53-year-old patient who had been admitted to the clinic with burning feet and restless leg syndrome associated with severe insomnia. Because of the severity of his symptoms, which had begun ten years ago, he had been on sick leave for a year. The symptoms eased when he took his shoes off and sprayed cold water on his feet. At night, he had to stick his feet out from under the covers. The tingling in his legs often felt like an electric current flowing up into his thighs. The leg restlessness woke him up. While working as a mechanic, in the end he could barely stand for three hours before he had to take his shoes off, cool his feet, and walk around. Additional diagnoses included arterial hypertension, varicose veins since age eighteen, venous thrombosis with venous ulcer at age thirty-four (treated by vein stripping), and back problems.

Patient History and Diagnostic Findings

The patient, a man of robust build, let his head hang limply and sadly. His closely spaced eyes wandered restlessly; he seemed almost incapable of eye contact. Tears came to his eyes halfway through his description of his symptoms. He appeared introverted, unsure of himself, and without hope. His legs were restless, and he immediately took off his shoes, explaining that he couldn't stand the heat and pins and needles in his legs. His memory with regard to his biography was conspicuously poor.

At age three, the patient had survived a life-threatening illness (no further details available). Since that time, he had suffered intermittently from asthma, although not at the moment. In elementary school, he lived in fear of a teacher who beat him, and he was also beaten by his father, who was a friend of that teacher. His grandfather, his most important confidant, died when the boy was nine. At age fourteen he began an apprenticeship with his father, who humiliated him in front of the other apprentices. Thanks to a good marriage and the birth of his son, the patient managed to extricate himself from his dysfunctional birth family, and five years later he also left his father's business.

In view of this situation, the patient seemed most in need of help with issues related to his biographical development and the unresolved crisis in his family; but we had our orders, and focused on finding treatments for his leg symptoms.

In terms of the integration of his constitutional members, we noted a strong, well-proportioned physical body. Early, possibly hereditary venous insufficiency suggested that the higher constitutional members have failed to take hold of the etheric body adequately. As a result, body fluids had become subject to gravity and the vessels lost their shape. The patient's wandering gaze, his high blood pressure, and the dysesthesia and restlessness in his legs were indicative of strongly one-sided astral activity. Instead of penetrating the etheric body from within, in the blood, his astral body was working externally on the skin, in the lungs, and in arterial circulation, which bears the imprint of the pole of consciousness. In the skin and lungs, the astral body was arousing pathological consciousness that manifested as leg pain and restless movements not controlled by the "I." The overriding phenomenon of insomnia indicated that the astral body was failing to fall asleep completely. The state of his "I" was also evident in his poor memory. Normally, the I-organization is incorporated into the lower part of the body around age nine.[91] In the case of this patient, that process had been severely encumbered by his anxiety-ridden relationship to his father and even more so by the loss of his grandfather at that age.

Finding a Treatment

In this case, finding a treatment followed directly from art. As we knew from working with the intervals (as astral body processes) in tone eurythmy, the gesture of the seventh manifests in excitation and restlessness of the limbs on the body's periphery. Describing the interval of the seventh, Rudolf Steiner said if people could experience it fully (something they usually avoid) it would feel as if their skin was peeling off; they would feel like the satyr Marsyas, who was flayed alive after losing a contest to Apollo.[92] We wondered, therefore, if a homeopathic preparation of Fluoricum acidum, the material seventh-process of the first period, might be able to help this patient.

According to Mezger, the homeopathic drug picture of Fluoricum acidum includes: "Compulsive movement, restlessness, varicosis; needs to stick the legs out of bed, better with cooling. Complementary to Silicea."[93] We were pleased both with the surprisingly apt drug picture and the remedy's complementarity to Silicea. Like carbon, silicon falls in the fourth group and serves structural development. Musically speaking, the fourth is the polar opposite of the seventh, as its eurythmy gesture also demonstrates.

The patient was prescribed Fluoricum acidum D30 (5 drops each night) to relieve him of the astral body's external seventh-process according to the "take-over" principle. Three months later, the patient reported feeling much better: he was able to sleep again, and from time to time he "forgot about his legs." During that period, he had one asthma attack that was treated conventionally. Comprehensive help for this patient would certainly require a comprehensive approach. In describing the path from tone eurythmy to homeopathic treatment, however, the point was simply to give an example of a "healing art" that unites a musical, Goethean understanding of substances with homeopathic empiricism.[95]

CHAPTER 3

The Experience of Music and Its Basis in Physiology

> *Thou muse of the art of sound,*
> *What inspirations you hold*
> *For deciphering the physiology*
> *Of the human soul!*
>
> JOHANN GOTTFRIED HERDER[96]

Experiencing Time in Music

At age thirteen or so, I had an experience I shall never forget: I saw and heard Wilhelm Backhaus perform in one of his last concerts in Stuttgart. With snow-white hair, he sat down at the black grand piano to play a Beethoven sonata. He laid his hands on the keys, and with the first note *the entire first movement* was present. This experience was completely unexpected, because no one had ever told me it was possible.[97]

To explore our musical experience of time more closely, let's take as an example the beginning of Beethoven's Piano Sonata, Opus 90 (see below). Drawing on the rhythm of the limbs, the movement begins deliberately in the minor key (measures 1 and 2); in recall and reflection, measures 3 and 4 turn to major. A heightened repetition (measures 5 to 8) starts in major, then turns to minor. After such exact rhythmic repetition of the first four measures, these four repetitions of the rhythmic motif demand something new; the repetition forms the "organ of expectation" for the new, which then appears in measures 9 to 16. Purely melodic two-part downward arcs in the major key transform the rhythm of the decisive, orchestral preliminary motif into a singing, gently descending movement. Finally, then, in measures 16 to 23, the

Example 3: Ludwig van Beethoven, Piano Sonata, Opus 90, first movement.

first motif appears anew, transformed into high, spacious, delicate structures. The compact upward fourth at the beginning (measures 1 and 2) is answered by a downward seventh in detached or relaxed surrender (measures 17 and 18). The initial rhythm, abandoned by the melody in the lyrical midsection, is preserved in the bass line. A relaxed flow of eighth notes brings the passage to a peacefully descending conclusion.

An ascending, compact and physical will-gesture is transformed in song, while its upbeat continues to pulse, and is ultimately transformed into a deeply downward-turning serene "understanding." After the final lyrical downward flow, the excess will of the opening measures breathes in matured "resignation."

The beginning breathes *future*; it comes from the limbs, from below. The middle sings the *present* exclusively. The illumined understanding of the third phase breathes *past*, conclusion. This is the path from the limbs through the midsection to the head: the human being as a fluidly sculpted time-body.

Music gains life from the encounter between the declamatory *melos* on the one hand and the pulsation of dance on the other; from dance and language. It originates in the confluence of cultic temple dance and sacred chant, as it is still present in Gregorian chant. In its highest form, music harmoniously unites the blood's will with the word's light, which shapes the stream of *melos* on the breath.

In the song-like, speech-like diction that makes a melody into a theme, time is shaped from the overarching perspective of consciousness of the whole.[98]

Musicians alternate between future-oriented development of new themes and the retrospective listening and recollection of "recapitulations," which also resound physically. New development is experienced at the hand of memory: music is observing time from within. The living basis of our experience of intervals is that one note persists within while the next note sounds. Conversely, the retrospective repetition of a new theme gives rise to increasingly willed demand for further development (see measure 8 above). Thus musicians continuously experience the "twofold flow of time" outlined in spiritual science.[99] The will to move, which lives in the blood, urges forward, while the astral body looks backward in the neurosensory process of hearing. The "I," which holds these two processes in balance, alternates between spirit presence and physical mastery of the instrument. It takes hold of the etheric stream in imagining what is to come and the astral stream in recalling the past (fig. 23).

We are made aware of the close relationship between the shaping of melodic movement and the movements of thinking by the phenomenological psychology of music of Ernst Kurth, Viktor Zuckerkandl, Jürgen Uhde, and Theodor W. Adorno.[100] The process indicated above lives in felt, willed, shape-giving *mental images* that are also reflected in outer sounds. All musicians are familiar with this "mental imaging" as the anticipation of their inner intentions. They experience thought-movements, submerged in feeling, in the presence of mind and spirit that exists between intentional anticipation and reflective recollection. Adorno says that the form-giving idea of the whole is at work in mastering time to the extent of maintaining a consistent tempo.[101] Musical

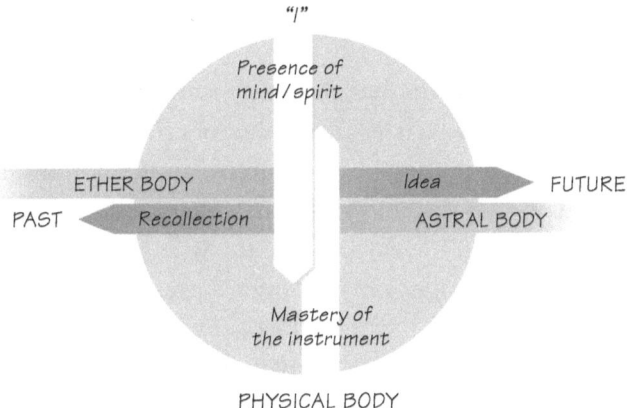

Fig. 23

feeling experiences the movement of mental images in the shaping will.

Rudolf Steiner's relationship to music was determined by this quality of musical movement, which is so closely related to the movement of thinking. Steiner reported in retrospect on the debate surrounding "program music," which had been enflamed by Richard Wagner, and which Steiner experienced at close hand in Vienna in the 1880s. "For me, thinking possessed *content* in and of itself, and did not receive it as a result of the perception that expresses it. This, however, leads as a matter of course to experiencing the pure musical sound-formation as such. For me, the world of musical sounds was the revelation of an essential aspect of reality. That music is still intended to 'express' something in addition to sound-formations, as Wagner's adherents were then claiming in all possible ways, seemed totally 'unmusical' to me."[102]

The thought-related aspect of musical experience might seem to contradict the will-character of musical movements in Schopenhauer's sense (presented in detail in Chapters 1 and 2). However, real thought-*activity* (as distinct from "having thoughts") is actually a process of willed concentration, just as musicians allow motifs to emerge out of the idea of the whole, step by step, one development after the next. In melodic feeling, the movement of ideas appears as illumined will, as Schopenhauer formulated it (see p. 22).

The Archetypal Music of Singing

All instrumentalists (whether they play wind instruments, strings, piano, or percussion) as well as all conductors *sing* inwardly to themselves along with the music they hear and perform. Even non-singers breathe as if singing as they listen.[103] This has been called *respiratoria auditoris*, "hearing breathing."[104] According to Wängler, people listening to recorded songs (that is, not in the singer's presence) breathe as the singer breathed. Roemer says that in subjects emotionally affected by music, respiration follows the music's rhythm, but no such connection emerges among those indifferent to music. "When the reproduction is perfect and the listener fully receptive, the artist's original respiratory rhythm will reappear in the listener."[105]

What Are the Physiological Foundations of Musical Perception of Time?

"If we listen attentively to a beating metronome, it is not at all difficult to give a subjective emphasis to every second (or third) beat. This is no longer possible, however, if the beats are separated by more than three seconds. Making the mental connection is possible only for a limited time. The impression of unity, the sense of connection between a perceived figure and the feeling of the immediate present ... is limited to a temporal window of three seconds."[106] In a different study, the same author writes: "Events are perceived not only in isolation but also in relationship to each other, so that successive events always form a single perceptual gestalt. This is possible because the brain makes a temporal integration mechanism available. A number of different examples illustrate this integration mechanism with its upper time limit of approximately three seconds.... For example, if we are interested in whether two sounds are equally loud or two lights equally bright, all experiments indicate that appropriate comparison is possible only when the two stimuli occur within a window of not more than three seconds. If the time interval between the two stimuli increases, the first stimulus fades and the second is overestimated.... A further question is whether the three-second limitation of this integration mechanism applies only to perception, or also to other areas of our experience and behavior."[107] Finally, Ernst Pöppel mentions that movement sequences are also planned "preferentially for three seconds." "If we want to execute a movement with great temporal precision" [think of musicians! A.H.] "the movement can only be planned about three seconds in advance."[108] This means that anyone playing a piece of music shapes time through the mysterious integration system that Pöppel investigated, which works in a three-second rhythm.[109]

Pöppel considered the brain the site of the "integration mechanism" that allows us to convert individual stimuli into a coherent perceptual gestalt: "This is possible only because the brain makes a temporal integration mechanism available." From the perspective of spiritual-scientific empiricism, Pöppel's "integration" is the formative thinking and forming of mental images described above in relationship to the experience of music.[110]

There can be no doubt that the "integration mechanism" is a brain activity. That is why the freedom of thinking that confronts musical interpreters with questions such as "Which notes belong together?" and "Does this note still belong to the previous motif or to the next one?" is also suited to this integration process.

But where does the three-second phase length of this integration process come from? Pöppel says only, "Nature has ensured that we can construct units, but only within limited temporal frameworks."[111] The next question should be, which organ of our "nature" has a rhythm with a phase length of approximately three seconds? Isn't it the frequency of *respiration*, which at eighteen breaths per minute has a phase length of approximately three seconds? Respiration transmits its rhythm to the brain and the sense organs via the cerebrospinal fluid. In works dating from 1916 to 1924, Rudolf Steiner identified the respiratory movement of cerebrospinal fluid as the integration process that underlies the experience of music and in fact all comprehension of integrative perceptions in art and thinking.[112]

According to Rudolf Steiner, our experience of music is based on the fact that respiratory movements extend into the inner ear and brain via the cerebrospinal fluid. Steiner says that respiratory movements of cerebrospinal fluid are the physiological prerequisite to all emotional activity, whereas neurosensory processes as such, communicate only the conceptual portion of the content of feeling.[113] In his book *Riddles of the Soul*, Steiner formulates it like this: "How does the emotional experience of music arise? The *mental image* of the sequence of notes, which depends on the organs of hearing and neurological processes, cannot account for it. The experience of music arises when the rhythm of respiration extends into the brain, where it meets the results of auditory and neurological activity. The soul does not live in what we hear or the mental images we make. *The soul lives in the rhythm of respiration; it experiences what is activated in respiratory rhythmic activity impacted by events in the nervous system.* If we simply see the physiology of respiratory rhythm in the right light, we will be able to acknowledge the full truth of the statement, '*experiences of feeling emerge when the soul relies on the rhythm of respiration, just as it relies on neurological processes in shaping mental images*'"[114] (emphasis added). A year later, in lectures to artists in Munich and Vienna in May and June of 1918, Steiner identified the respiratory movements of cerebrospinal fluid as the connection between respiration and the neurosensory system.[115] On this basis, he went on to develop a comprehensive physiology of artistically creative imagination.[116]

History of Research into the Respiratory Dynamics of Cerebrospinal Fluid

With his theory of the respiratory dynamic of feeling, Rudolf Steiner definitely fell in line with the natural-scientific approaches of his time. At the latest, by the time of the publication of Hyrtl's *Lehrbuch der Anatomie des Menschen* (tenth edition, Vienna 1867), the respiratory rhythms and the pulsation of cerebrospinal fluid and the brain belonged to the basic knowledge of medicine. As early as 1811, the movement of cerebrospinal fluid in response to respiration was demonstrated experimentally (possibly for the first time).[117] Subsequently, a plethysmograph was used to record respiratory and pulse-dependent brain movements in patients with brain injuries and cranial defects; and the effects of arithmetic problems, sensory stimuli, and emotional excitement on brain movements were studied. Both Mosso (1881) and Resnikow and Dawidenkow (1911) found that brain movements were more significantly affected by emotions than by intellectual activity. In their 1951 study *Über das Verhalten des Liquordrucks bei psychischen Vorgängen*, Bener, Kehrer, and Knebel determined that cerebrospinal fluid pressure increased as more difficult arithmetic problems were assigned.[118]

In modern research into the physiology of emotions, this line of investigation appears to have been forgotten. Authors such as Oliver Sacks,[119] Antonio Damasio,[120] and Manfred Spitzer[121] consider only the neurometabolism and blood supply of brain tissue.[122] At least Spitzer admits, "To date there have been very few studies of exactly what happens in the brain when it (sic!) responds emotionally to music."[123]

In the anthroposophic medical literature, Eugen Kolisko was the first to take up Rudolf Steiner's ideas. His afterword to Valborg Werbeck-Svärdstöm's singing textbook includes a first Goethean physiology of the cerebrospinal fluid as it relates to respiration in singing and hearing.[124] Gisbert Husemann applied Goethean methodology in his detailed descriptions of the anatomical, physiological, and psychological aspects of the cerebrospinal fluid system with respect to buoyancy.[125] In 1989, the author of this book then followed up on these studies by outlining a musical physiology of the "formation and flow of cerebrospinal fluid."[126] Respiratory movement in its significance for musical consciousness is not touched upon in that work, which deals only with the embryonic development of the cerebrospinal fluid system and fluid circulation in general. The connection of that work to the material presented here will be clarified at the end of this book.[127]

How Does Respiration Move the Cerebrospinal Fluid?

The Cerebrospinal fluid is produced inside the ventricles of the brain and it flows outward through the three openings in the area of the cerebellum to join the cerebrospinal fluid around the outside of the brain

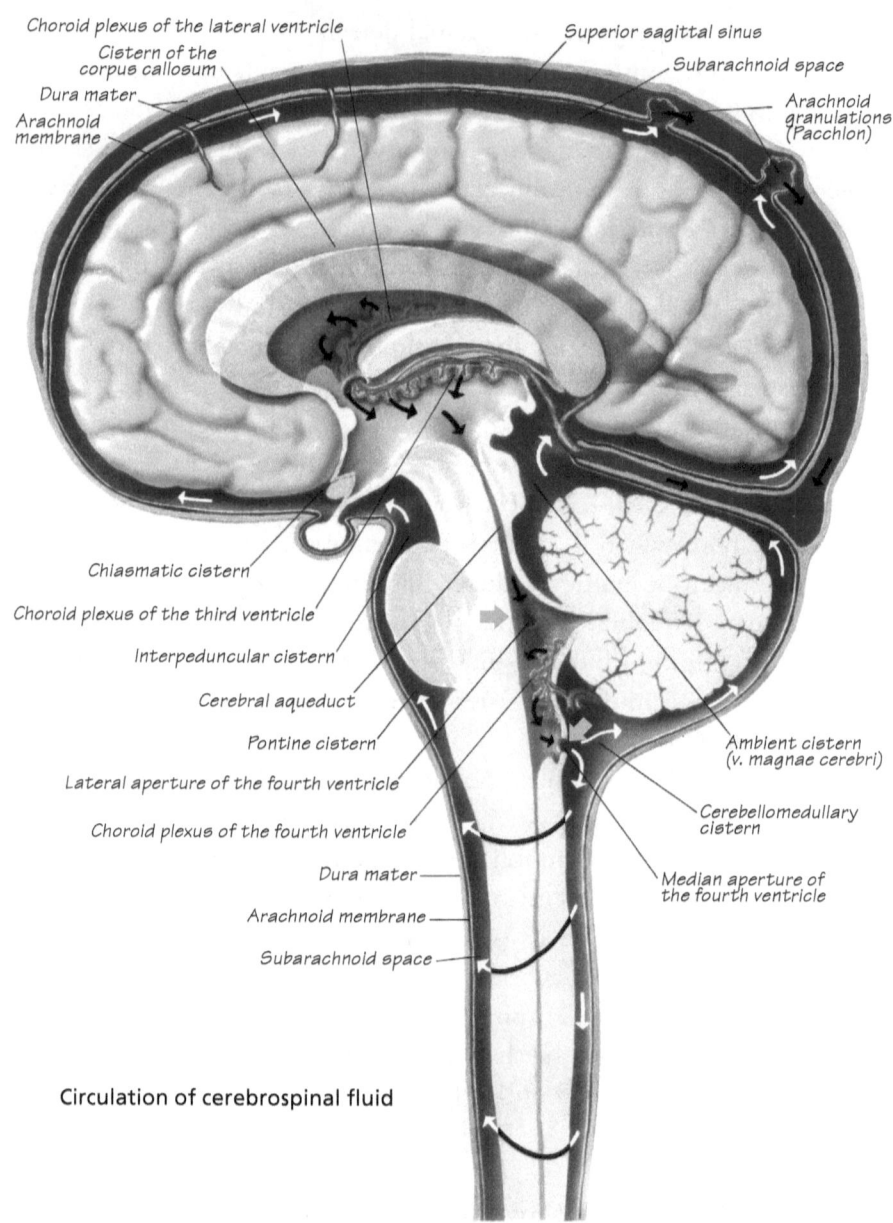

Circulation of cerebrospinal fluid

Fig. 24: The cerebrospinal fluid system. Black arrows = internal cerebrospinal fluid; white arrows = external cerebrospinal fluid; grey arrows = holes through which the internal cerebrospinal fluid flows outward.

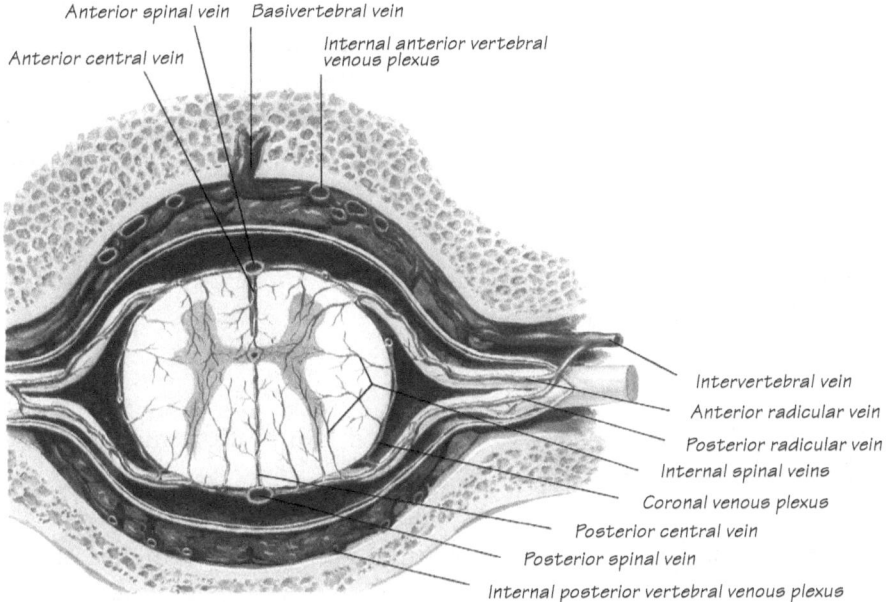

Fig. 25: Cross-section through the spinal cord.

(fig. 24). The brain floats in the outer layer of fluid. According to Archimedes' principle, it is buoyed up by a force equal to the weight of the fluid it displaces, so instead of weighing approximately 1,300 grams, the brain weighs about 30 grams.[128] The dura mater, the tough outer layer of the meninges, lies against the bone of the skull, but in the spinal canal it forms the *dural sac*, which is not attached to the surrounding bone. The spinal cord, floating in cerebrospinal fluid, is contained within the dural sac. The epidural space between the dural sac and the mobile walls of the vertebral canal is filled with adipose tissue that contains a network of veins. These veins are valveless, which allows free upward and downward flow of venous blood during respiration.

In infants, respiration-dependent movement of the cerebrospinal fluid can be seen and felt in the larger (anterior) fontanelle. When the baby cries, the fontanelle bulges. When inflammation of the brain or its meninges is suspected (encephalitis or meningitis), fluid is extracted from the dural sac for diagnostic purposes. When diagnostic lumbar puncture is performed on a crying infant (and unfortunately, given the need for the nurse's secure grip, crying is likely even before the puncture is performed) pressurized fluid spurts from the puncture needle in a high arc with every cry, but the pressure drops visibly and rapidly whenever the baby pauses for breath.

These clinical experiences clearly teach us that respiration regulates the pressure and movement of cerebrospinal fluid. But how is respiratory movement transmitted to the fluid?

Abdominal Veins as Mediators between Respiration and Cerebrospinal Fluid

During inhalation, the diaphragm sinks, reducing the volume of the abdominal cavity. But since the abdominal cavity is essentially filled with incompressible fluid, a corresponding volume of blood is forced out of the abdominal venous plexus and into the veins of the spinal canal via many different connections.[129] Pressure tests reveal synchronous transmission of pressure between the extradural veins and the cerebrospinal fluid.[130] Because the system of spinal veins is valveless, venous blood can rise and descend freely during inhalation and exhalation, just as the cerebrospinal fluid does (fig. 26).

Schroth as well as Klose and Winkler characterize the "epidural venous plexuses" as the transmission system ("compliance system") for the movement of cerebrospinal fluid in the spinal canal and in the brain. In the process, up to 15 ml of fluid are displaced.[131]

The Extension of Respiration into the Interior of the Cerebral Ventricles

Respiratory movement also influences the movements of fluids in the interior of the cerebral ventricles in different ways: As the pressure of the outer fluid increases, the flow from the cerebral ventricles slows, stops,

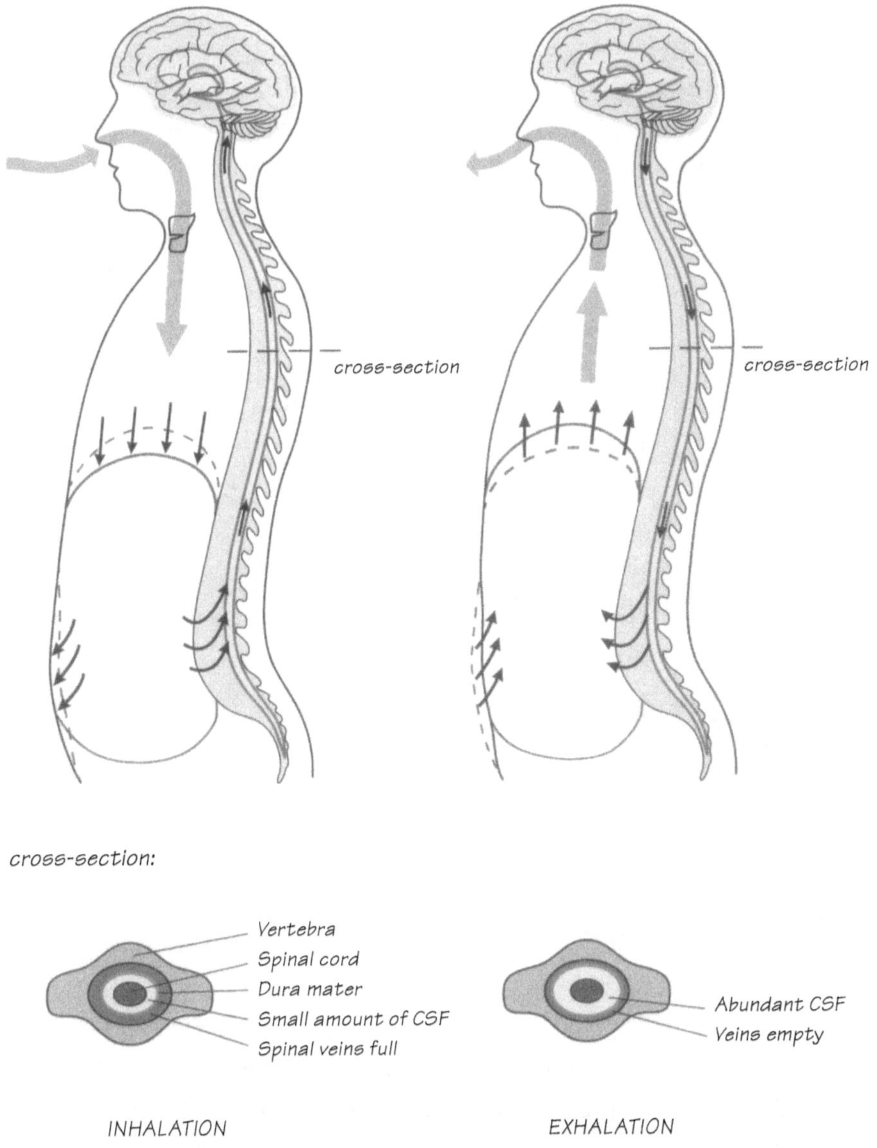

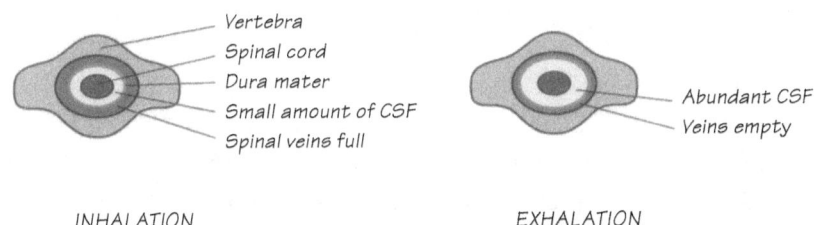

Fig. 26: Respiratory movement of cerebrospinal fluid (CSF), see text (Illustration by Edgar Bayer).

or is reversed, flowing back into the ventricles. As the pressure drops, the flow from the ventricles accelerates. Increasing pressure is most commonly a consequence of inhalation but may also occur during crying, singing, or speaking (that is, percussive or plosive consonants, such as *d, t, k, g*). Decreasing pressure is first and foremost a result of relaxed exhalation.[132] Rudolf Steiner states that cerebrospinal fluid moves up into the skull during inhalation and downward during exhalation.[133] This phenomenon has been repeatedly confirmed in abdominal respiration.[134]

Respiratory movement contrasts with the much more subtle cerebrospinal fluid (CSF) movements with which arterial pulsation modulates flow from the ventricles. During arterial systole, the increase in volume of the brain together with its vessels results in slight compression of the ventricles in places. This compression, however, amounts to only approximately 0.5 ml per heartbeat or 2 ml per respiratory cycle.[135] CSF, therefore, is the medium through which the activity of the lungs and heart extends into the brain.

Respiratory Dynamics, the Inner Ear, and the Experience of Music

According to Steiner, the experience of music arises because "the respiratory rhythm in its extension into this organ [of hearing] encounters what is accomplished by the ear and the nervous system" (see pp. 57-58). The inner ear and the organ of balance float within bony canals filled with fluid (perilymph). Through the cochlear aqueduct (also called the perilymphatic duct), the perilymph extends into the CSF-filled space of the skull. The endolymph, the fluid within the cochlea, in which the auditory waves "break" during hearing (see fig. 8, p. 9), also transmits waves of pressure through the endolymphatic duct into the fluid-filled interior of the skull (see fig. 7, p. 8).

Today the hydrodynamic connections between the fluid-filled spaces of the inner ear and the CSF-filled spaces of the brain are uncontested. "Intracranial CSF pressure can be transmitted directly to the fluid of the labyrinth."[136] Pressure is transmitted primarily via the perilymphatic duct but also via the endolymphatic duct.[137]

Meanwhile, therefore, the anatomical and physiological prerequisites upon which Rudolf Steiner based his explanation of the experience of music have been confirmed by natural-scientific findings. Steiner was not able to draw on the research of his contemporaries, interactions between CSF pressure and the inner ear had not yet been proven at that time. Scientific confirmation of his statements, however, has become available in retrospect.[138]

Breathing between Life and Consciousness

Cerebrospinal fluid is produced in the choroid plexuses located in the interior of the brain on the floor of the four ventricles (fig. 24, p. 60). These organs convert arterial blood into CSF on the one hand and venous blood on the other.[139]

When held up to the light, the CSF of healthy individuals is as clear as water. It contains none of the components that make blood a life-giving substance that supplies the metabolism. It lacks red and white blood cells, and the proteins and nutrients have been removed. Blood proteins are present only in microscopic quantities, their concentration reduced by a thousandfold. Glucose and sodium chloride are the only substances present in about the same concentrations in blood and CSF.[140]

How the Cerebrospinal Fluid Moves with Respiration

The movement of both *venous blood* and *cerebrospinal fluid* depends on respiration. A vein stands out on an angry person's forehead because blocked exhalation has brought venous flow to a standstill throughout the body. In contrast, *arterial* blood flow is independent of respiratory movement. Given the vital importance of arterial perfusion, the possibility of its being interrupted by every shock, fit of rage, or other reason for holding the breath (as is the case with venous circulation) would be incompatible with life itself.[139]

When we are frightened, consciousness impacts the movement of blood directly, via the neurosensory system. Venous blood stands still, and for a moment the cerebrospinal fluid resembles a "transparent crystal." The same dynamic occurs when we concentrate intently while solving a difficult arithmetic problem, for example. Try this experiment for yourself: observe what happens to your breathing as you try to calculate "17 x 29 = ?" in your head.

When thought activity intensifies, respiratory movement slows or is interrupted (see p. 64). As a result, venous flow from the brain slows or stops, breakdown products such as carbon dioxide accumulate, and tissue metabolism becomes acidic. In other words, holding one's breath entails functional intensification of the brain's deposition tendency, the salt process (see chapter 2, p. 47).

There are a number of different Goethean and spiritual-scientific perspectives on the qualitative aspects of arterial and venous blood. Physicians know that an arterial blockage means tissue death in the affected organ. As physicians and caregivers, we see the patient's feet turning first *white*, then *blue*, then *black*, and we think of a heart attack. Blockages in venous flow seldom have such direct vital consequences. This is the aspect Rudolf Steiner had in mind when he associated arterial blood with *life* and *anabolic processes* and venous blood (with regard to the metabolic end products of endogenous catabolism) with *death* and *catabolic processes*. Through inhalation, cosmic life forces work in arterial blood, while earthly, catabolic death forces work in venous blood through the metabolism.[141] In arterial blood, we must imagine the etheric and soul-spiritual aspects of the human constitution as being more completely absorbed into the physical body. In venous blood, the etheric body is more loosely connected to the physical body in order to allow the connection to dissolve during exhalation.

In the way respiration shapes and dominates the movement of venous blood and CSF, we see the action of the astral body, which is at home in air, as it intervenes in physical and etheric processes.[142] At this point, making the step to imaginative image-forming becomes necessary.

An Imaginative Physiology of the Brain's Respiratory Dynamic

"The sculptor of the physical body" is an image that Rudolf Steiner often used to describe the etheric body. Far from being a mere metaphor, this image guides our thinking into the imaginative reality of etheric formative forces.

Let's imagine a sculptor, such as Michelangelo, at the height of his creativity, with a large studio, many employees, and much demand for his work. Chisels, always freshly sharpened by apprentices, hang ready for use; trusted workers oversee the transportation of blocks of marble from Carrara to Florence. We can take this scene as a metaphor for the etheric body in arterial blood. Everything the sculptor needs is always in place: nutrients, oxygen, new erythrocytes and other blood cells. Employees remove any waste; everything is fresh and ready for the sculptor, who breathes in images from the cosmos and is eager to create the forms of physical organs. But then something unheard-of happens. One morning, Michelangelo enters his workshop and finds it empty! Sculptures, stones, tools gone—thieves have stolen everything in the night! This is the situation the etheric body confronts when it flows out of the arterial blood through the choroid plexus and into the cerebral fluid. Nothing that would allow it to build up the physical body is available anymore. In the cerebral fluid, the etheric body comes up empty.

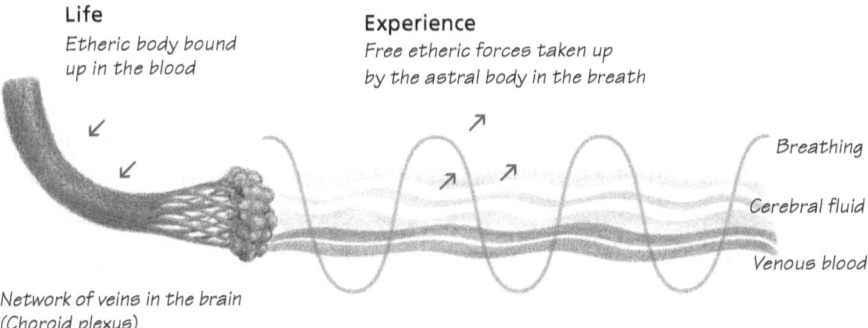

Fig. 27: The metamorphosis of formative forces in the formation and movement of CSF. Arterial flow depends mainly on the metabolism and the heart. Venous flow and CSF movement depend primarily on respiration and consciousness (see text) (Illustration by Edgar Bayer).

At that moment, however, its movement is taken up by the *astral body* in respiration. In the astral body's breathing, the etheric body awakens to consciousness and becomes consciously felt, *living experiencing (Erleben)*.

The astral body itself sucks all the vitality out of the blood in the choroid plexus. It extracts the life from the physical body, breathing it in as feeling-experience, as an inner world of images. The astral body wants to feel life. Here, free of the physical body, the "I" within in the astral body is also free in the forms it wants to give to life in images, thoughts, and sounds. Here we see the physiological basis for the freedom to shape images in thinking. Here formative forces are freely available, wrested away from the physical body in the head.

To the extent that the etheric body submits to the astral body and the "I," it extricates itself from physical forces. This phenomenon can be seen in the reduction in the brain's weight (from 1,300g to 30g) as the result of buoyancy, according to Archimedes' principle. "This high degree of deactivation [of matter] allows the etheric body to become effective on behalf of the brain to an exceptional extent. [The etheric body] *can do what it wants* because it is not misled by matter's heaviness."[143] At the beginning of this book, we already had the opportunity to observe this behavior of the etheric body in the sensory activity of the eye and ear (p. 2 ff.).

Respiration as the Organ of Creative Thinking

In the formation and flow of CSF, the astral body and "I," together with the etheric body, separate from the physical body. This is the same process that the entire human being undergoes in crossing the threshold of death. At the moment of death, the etheric body, astral body, and "I" remain united, and the physical body separates from them. The awakening we experience then, however, reaches greater depths of existence, because in the moment of death the lungs, the heart, and the entire will-system of metabolic organs become "brains," so to speak. As the dying body releases life, we awaken in all aspects of life in which sleep and forgetting formerly had prevailed.

Fig. 28: Lateral view of a casting of the CSF-filled ventricles (left = front). Drawing (from a model) by Daniel Moreau. (See fig. 24, p. 60).

In the uniquely shaped hollow, bird-shaped space (fig. 28) that reflects the buoyancy of the cerebral fluid, we live in creative forces that we extract from the body.

Here lies the secret of "death in the midst of life," to which we owe the waking consciousness of our thinking and the experience of creative formative forces resurrected from the dead. Goethe spent a lifetime practicing to develop these forces of creative imagination into the organ of cognition for the life of the organic world. In old age he achieved an inspiration of life in death when he contemplated Schiller's skull:

Within a gloomy charnel-house one day
I view'd the countless skulls, so strangely mated,
And of old times I thought, that now were grey.

…

Vainly ye sought the tomb for rest when tired;
Peace in the grave may not be yours; ye're driven
Back into daylight by a force inspired;
But none can love the wither'd husk, though even
A glorious noble kernel it contained.
To me, an adept, was the writing given
Which not to all its holy sense explained,
When 'mid the crowd, their icy shadows flinging,
I saw a form, that glorious still remained.
And even there, where mould and damp were clinging,
Gave me a blest, a rapture-fraught emotion,
As though from death a living fount were springing.
What mystic joy I felt! What rapt devotion!
That form, how pregnant with a godlike trace!
A look, how did it whirl me tow'rd that ocean
Whose rolling billows mightier shapes embrace!

(From "Lines on Seeing Schiller's Skull" by J.W. Goethe)

"As though from death a living fount were springing.... A look, how did it whirl me tow'rd that ocean Whose rolling billows mightier shapes embrace!" Imaginations of the etheric world are revealed to the poet—the spiritual sea of life from which shapes come flooding forth. Here poetic inspiration transported Goethe into the second region of spirit land, where the creative archetypal images of life are found, as described in Rudolf Steiner's *Theosophy*: "... but here this life forms a perfect unity. It streams through the spiritual world like a fluid element.... This second level of the country of spirit beings might be described as 'flowing life formed from thought substance.' ... As long as we are observing the world from the vantage point of physical embodiment, life appears bound to individual living things. In the country of spirit beings, however, life is released from individual things and flows throughout like the blood of life, as a living unity that is present in all things."[144]

In his natural-scientific studies, Goethe developed the first stage of this imaginative perception of life—the organic "type" that manifests physically in various metamorphoses. On multiple occasions, Rudolf Steiner referred to this inwardly living idea (for example, the archetypal plant) as a "fluid idea" because to organic perception, it "flows" into physical metamorphoses, so to speak.[145] According to Steiner, the Goethean "type" can no longer be grasped with brain-bound thinking but only through the *respiratory movement of cerebrospinal fluid*, the instrument of creative consciousness. "Of course abstract thoughts are still bound to nerve activity, but pictorial thinking is also bound to the activity of respiration. We can say, 'Here is formative life.' Inasmuch as we breathe, we have formative life in us."[146]

Breathing Colors and Sounds

Through respiratory oscillations, scientists can call thinking to life, and artists can enliven sensory processes.[147] When the CSF rises during inhalation, it flows forward in the dural sheath of the visual nerves to the eyes.[148] During exhalation, it flows back, away from the eyes into the interior. Without "breathing" with our eyes, we see "red"; we register the color. When we "breathe" red, we awaken to its sensory and psychological qualities. We experience how it comes toward us; it becomes living activity. With blue, we experience how it recedes, moving away from us into the distance (see Goethe's *Theory of Color*, part 6, § 758 ff.) On the feeling level, we participate in the soul-dynamic of color when respiration, stimulated by nerve action, meets the metabolic blood process in the retina.

Something comparable is also true of the ear. The ear and the brain allow us to distinguish different pitches of notes, but when we *breathe* them (meaning that the movement of CSF continues into the inner ear via the cochlear aqueduct and the endolymphatic duct) [fig. 7, p. 8] we are conscious of the melodic gesture of the interval between the notes as a third, a fourth, and so forth. In the movement gestures of the intervals, musicians experience the dynamic of the sound ether from the perspective of the astral body's perception.

Instrumentalists and conductors "sing along" with their inner impression of the melody. Enthusiastic aficionados also sing to themselves

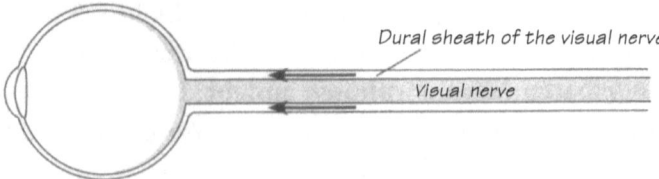

CSF flows toward the eye during inhalation

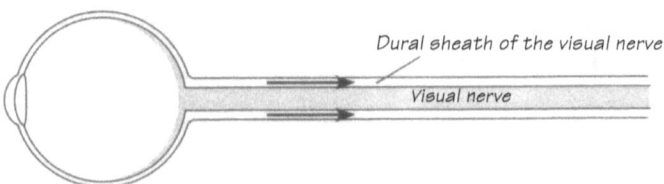

CSF flows away from the eye during exhalation

Fig. 29

as they listen, involving their larynx and their breathing in the hearing process. In the larynx, they configure the freed life forces in the CSF in ways that instrumentalists would then transmit to their instruments via the limbs. In the respiratory movements of the CSF, therefore, the *formed* sound ether coming from the outer world via the ear meets and mingles with the *formative* sound ether emerging from the larynx. The sensory perceptive process and the will activity interact in breathing.[149]

Legato and Staccato – Articulation and Phrasing through Breathing

The following example from Beethoven's violin concerto demonstrates how breathing can shape the phrasing of a melody.

As the violinist "sings" these four measures silently in two phrases of two measures each, as marked, the inner ear and the CSF interact as follows: as the instrumentalist inhales, CSF rises rapidly. Through inner singing, the

The Experience of Music and Its Basis in Physiology 73

Example 4: Ludwig van Beethoven, Violin Concerto in D major, Opus 61.

musician sends inner formative impulses of his/her musical imagination out from the larynx to meet the sounds reaching the ear (fig. 30).

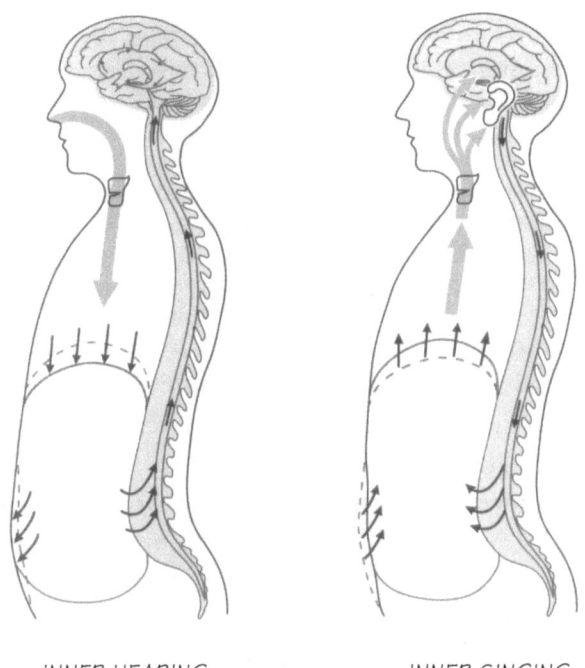

INNER HEARING   INNER SINGING

Fig. 30

Fig. 31: Sphinx from the temple of Aphaia at Aegena. (Photo of a replica by Marco Bindelli).

As a result, the seven breaking waves of each phrase, stimulated by the seven notes played on the violin, are subsumed into a single upward flow of CSF, as if in a common etheric body. Played legato, they become a structured, living organism. Ideally (that is, after sufficient practice), what the musician hears from outside will exactly match what was "sung" internally. The streaming sound ether does not simply die away into a mere perception of sound, as happens in non-musical hearing in the brain's neural processes. Rather, this sound ether is shaped by inner singing in the freely streaming ether of CSF respiration. Here it is the musician whose "rolling billows mightier shapes embrace," as Goethe put it (p. 70). Figure 31 shows a Greek sculptor's imaginative carving of this etheric head process—a sculptural image of musical hearing.

We *sculpt* the flow of a melody when we phrase and articulate it. In this process, the head-like formative force of the larynx in the sound ether is readily observable. Two polar formative principles here are "legato" and "staccato." In sculptural terms, the violinist in the previous example from Beethoven's violin concerto shapes two self-contained, living surfaces that arch in opposite directions (legato). A staccato passage, in contrast, isolates projecting sculptural elements, as in the following example.

Example 5: Ludwig van Beethoven, Piano Sonata Opus 14, no. 2: Andante.

How does breathing shape a staccato motif such as this? The breath is held after each note. The respiratory CSF wave stands still each time a travelling wave breaks in the inner ear. The CSF wave is patterned after the wave in the inner ear. In comparison to legato, which makes us somewhat dreamy, staccato wakes us up. This effect is based on the fact that breathing (and thus also CSF movement) stands still after every note. On the qualitative level, the astral body experiences the cessation of physical sound between notes as "death moments" that enhance its wakefulness. In staccato passages, the breath follows individual sound events in the ear without creating unity among them, as it does in legato passages. Instead, it conforms to and becomes one with neurosensory activity. In the legato experience, the astral body submerges into the etheric body and a totality results. In staccato, the astral body shifts upward into sensory activity and awakens as the music breaks up into isolated notes.

This is how we initially experience the shaping of melody, the uppermost domain of music. "The actual melodic element is readily comparable to the sculptural element. The sculptural element is arranged spatially, while the melodic element is arranged temporally, but anyone with an active feeling for melody's temporal orientation will soon realize that it contains a 'temporal sculpture' of sorts."[150]

Musical "breathing," however, can also free itself from its physical foundations and become active in the body-free state of pure soul. The element that is initially anchored in the physical and etheric bodies in song-like sculpting is then applied independent of physical breathing; for example, when a director or pianist uses legato and staccato at the same time, as in the following passage.

Example 6: Ludwig van Beethoven, Piano Sonata Opus 28, second movement.

To give additional examples, when a pianist or organist plays a fugue, several "songs," each at a different stage of breathing, may be shaped simultaneously. In this process, polyphonic music is also drawn into the sphere of thinking. Or when a staccato passage is played at a speed that physical breathing cannot replicate, the "I" within the formative sentient soul grasps only the superordinate phrasing of the passage on a pure soul level, in gestures that the astral body has "learned" from song-like breathing.

Dionysus, or Breathing Downward

In musicians, breathing (as a process of the middle and of balance) must extend as far *downward* as it does *upward*. In the lower body, breathing is asleep. It submerges into the anabolic, physiologically creative "music" that attunes the harmonies of the hormonal balances that carry out protein synthesis and breakdown in the liver. This spiritual music, which is active in the metabolism of fluid moon forces, also shaped the embryonic body. "Before musicians allow their music to resound in the world, the cosmic being of music has taken hold of them, filling them with cosmic music and incorporating it into them."[151]

How does this inner, spiritually creative music in the lower human body awaken in musical breathing? It does not live in Apollonian melody here, in the light and in the diction of speech; instead, it rises urgently in changing harmonies and rhythms in which the Dionysian life-processes of growth and renewal of reproduction strive to assume form in the soul's dynamic. This is the world from which composers draw and create new music, even if they are deaf, like Beethoven. Let's examine the melodic introduction to Chopin's famous Etude in E major, Opus 10, no. 3.

Example 7: Frédéric Chopin, Etude Opus 10, no. 3, measures 1-8.

In measures 9-13, the reflective repetition of this self-contained theme issues a challenge to Dionysus, who is then caught by Apollo in a three-measure passage ("crescendo," "stretto," "con forza") culminating in the victorious Apollonian G-sharp chord, which radiates peaceful light throughout the descending melody of the next five measures.

Example 8: Frédéric Chopin, Etude Opus 10, no. 3, measures 14-20.

In the middle section of the piece, we descend into a world that Apollo's light cannot penetrate. Only the rhythm still imposes any form on the passionate struggles in which we experience will-impulses being born out of the transforming fire of substances. In earlier times, these stormy tritones, surging both upward and downward, would have been perceived as the screams of Sorat, the Sun-demon banished to the underworld.

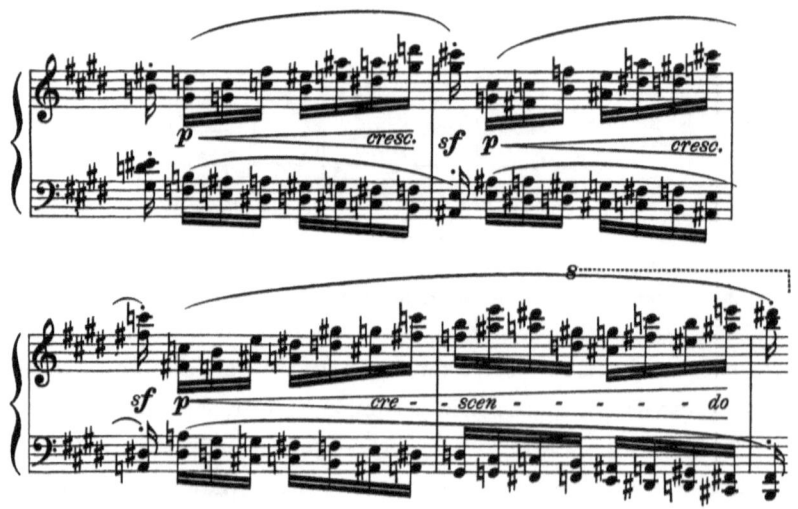

Example 9: Frédéric Chopin, Etude Opus 10, no. 3, measures 38-41.

This chaotic world is then confronted by the power of balance, the power of the "I," which leads the chromatics downward in wonderfully clear sixths, infusing the chaos with glowing, formative light.

Example 10: Frédéric Chopin, Etude Opus 10, no. 3, measures 48-53.

In alchemical terms (see chapter 2), an excessive, aggressive sulfur process from below is met from above by a healing, energizing phosphoric light combined with iron radiations.

Inhalation and Musical Inspiration

In the first chapter, we explored the "ear nature" of the lower human body. We will understand this aspect of hearing more precisely when we compare the route of inhaled air with the air's route into the ear.

As we follow the flow of air from the larynx through the windpipe into the lungs, we are tracing its *mechanical* route to the diaphragm rather than its *chemical* route into the blood. This is comparable to the route sound follows in the auditory canal until it reaches the eardrum, and the diaphragm's movement is analogous to the eardrum's vibration. Here is an overview of the correspondence between the "upper ear" and the "lower ear":

Hearing with the Head

outer ear → auditory canal → eardrum → middle ear → oval window → traveling wave → hair cells in the inner ear

---

larynx → windpipe and lungs → diaphragm → venous blood in the abdominal cavity = "compliance system" → dural sac of the spinal cord → wave in external CSF → CSF-contacting neurons in the interior of the cerebral ventricles

The Hearing Process in Inhalation

On the other side of the diaphragm, venous blood transmits inhalation to the CSF of the spinal cord just as the auditory ossicles transmit sound to the lymph of the inner ear. The venous system is the "compliance system" between respiratory motion and CSF movement (see pp. 62 ff.).[152]

CSF-Contacting Neurons

Without developing any details themselves, Eugen Kolisko (1938) and Gisbert Husemann (1984) suggested that such a metamorphosis of the ear does indeed exist.[153] Pursuing their suggestions enables us to understand the initially puzzling CSF-contacting neurons discovered by recent research. CSF-contacting neurons are *sensory cells* with afferent neurons. Many different types of these cells have been discovered in the ependyma (the membrane lining the ventricles of the brain) (see figs. 32 and 33). On the basis of its microanatomy, one type of cell, with hairlike processes extending into the inner CSF, was compared by its discoverers to the hair cells of the inner ear. Cells of this type, so-called mechanoreceptors (see pp. 8 ff. and p. 12), *can serve no other function than responding to CSF movement*, but to date no one has asked what significance this perception might hold for consciousness.[154]

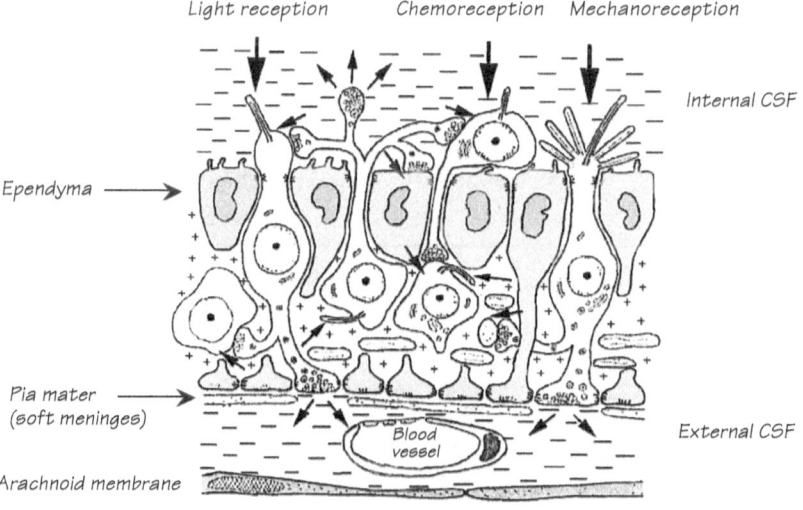

Fig. 32: Diagram of the different types of CSF-contacting neurons and their connection to the internal and external CSF and intercellular fluid. The arrows indicate non-synaptic, synaptic, and neurohumeral signal transmission. From B. Vigh, M. J. Manzano e Silva, C. L. Frank, C. Vincze, S. J. Czirok, A. Szabo, A. Lukáts, and A. Szél, "The System of the Cerebrospinal Fluid-Contacting Neurons. Its Supposed Role in the Nonsynaptic Signal Transmission of the Brain." *Histology and Histopathology* 19: 607-628 (2004). Reprinted with kind permission of the authors.

The Experience of Music and Its Basis in Physiology 81

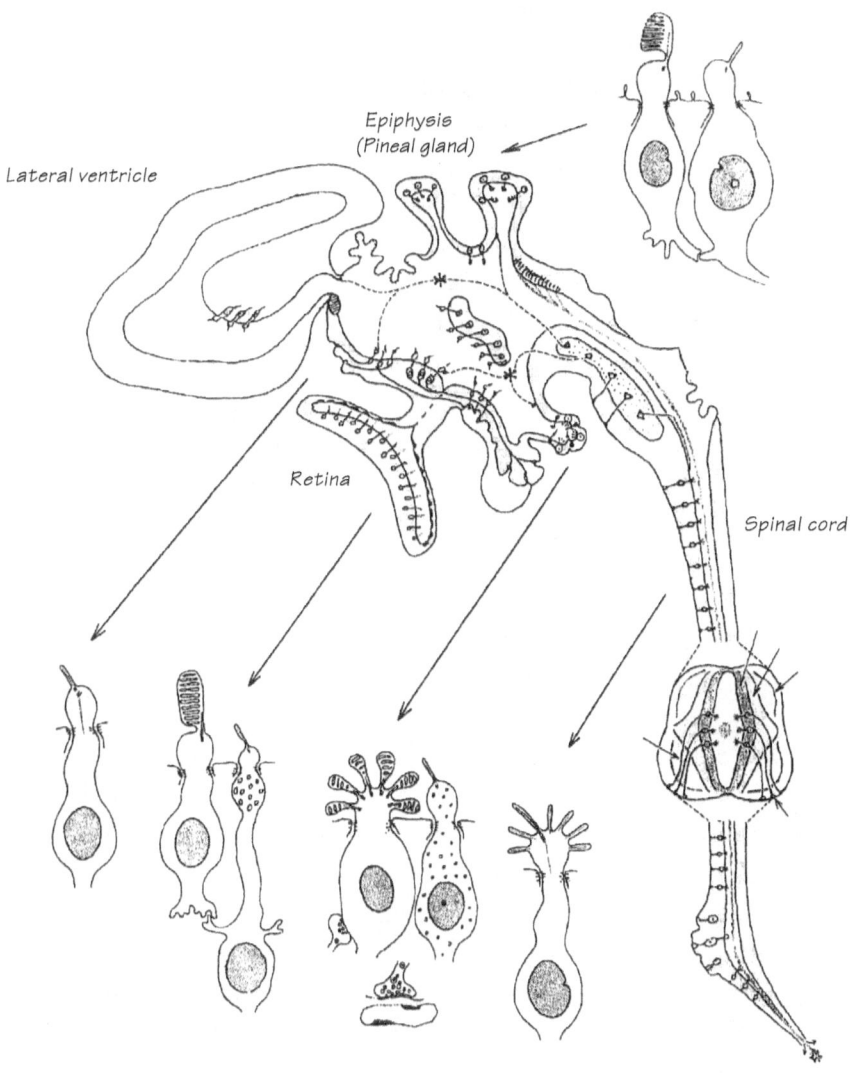

Fig. 33: Diagram of the system of CSF-contacting neurons and their different cell types. Often found in addition to chemoreceptors and light receptors are so-called mechanoreceptors, which react to flow movement in the CSF. From Béla Vigh and Ingeborg Vigh-Teichmann, "Actual Problems of the Cerebrospinal Fluid-Contacting Neurons." *Microscopy Research and Technique* 41: 57083 (1998). Reprinted with kind permission of the authors.

The movement of internal CSF in the ventricles of the brain is affected when respiration briefly allows the external fluid to flow *back* into the ventricles, as in singing (with or without sound) or in speaking (plosive consonants, for example) (see p. 64). Thus, this reversal of flow, along with the arrest or delay of flow out of the ventricles, in compliance with "hearing-breathing" (Höratmung) when making music, is mediated by the "inner ear" of the brain ventricles.

In the ventricles, therefore, the fluid movements of inhalation and exhalation that are produced by the larynx during internalized singing while experiencing music are conveyed to consciousness by CSF-contacting neurons in a metamorphosed inner ear process.

Rembrandt illustrates the inward larynx activity of inspired hearing in his portrait of the apostle Matthew, who "grasps" the inspiration of his angel with his larynx (fig. 34).[155]

Apollo's Lyre – from Chemical Process through Respiration to Light

When the metabolically-based venous blood of the abdominal cavity transmits the impulse of respiration to the CSF, the etheric body flows out of the blood stream into the wave of CSF, where it frees itself from the physical body as it does in the cerebral fluid of the brain itself (see pp. 67 ff.). In both instances, the chemical ether that is purposefully creative in the metabolism is freed from the blood; completely so in the brain, partially so (in dreaming consciousness) in the abdominal cavity. Seized by the movements of breathing, it streams into the inner light of neurological processes to become the *sound ether*, since the life process active in respiration is the *light ether* (see chapter 2).

When respiration takes hold of the chemical ether, it is displaced out of its own fluid element into light ($y \rightarrow x'$ instead of $y \rightarrow y'$).[156] This is the birth of *inner music* in the human being—the creative source of new music for composers.

In the imaginative terminology of spiritual researchers, therefore, the musical dynamic of respiration, as described here, is called "Apollo's lyre."

Fig. 34: Rembrandt van Rijn, Evangelist Matthew.

84   HUMAN HEARING AND THE REALITY OF MUSIC

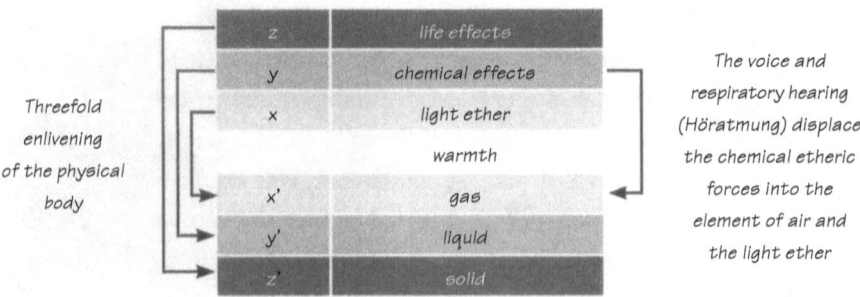

Fig. 35

Through it, Dionysus awakens in light. *Musicians who shape music bring the inner music of their own body, which is freed up in inhalation, on "Apollo's lyre," up to the head, where it meets the outer sounds perceived by the ear.*

To painters, musicians, and poets in Vienna and Munich in 1918 (Rainer Maria Rilke and Albert Steffen were in the audience), Rudolf Steiner elaborated as follows on the account he had published one year earlier. "We believe that the ear and perhaps the brain and nervous system are involved in the pleasure of music, but that is only a very superficial view. Physiology is only in its earliest beginnings in this area, and this aspect of physiology and biology will achieve a certain peak only when artistic thoughts flow into it. The foundation we are seeking here is something very different from mere auditory or neurological processes. The foundation of our perception of music can be described as follows: Every exhalation causes the brain (the skull cavity, the interior space of the head) to allow its cerebrospinal fluid to descend through the dural sac of the spine to the area around the diaphragm. Exhaling causes a descent. The opposite process is associated with inhalation: the cerebral fluid is driven back up toward the brain. A rhythmic ebb and flow of cerebrospinal fluid is constantly taking place within the arachnoid membrane, through broader and narrower spaces that are more or less elastic. The sounds that stream in through the ear, the impressions of sound that live in us, *become music when they encounter the inner music that results from the fact that the entire organism is a remarkable instrument, as I have just described.*

If I were to describe everything to you, I would describe a wonderful music within the human body; music that is not heard, but is nonetheless experienced inwardly. *What we experience musically is essentially nothing other than outer music meeting with the inner singing of the human organism.* With regard to what I just described, this human organism is a reflection of the macrocosm, which we carry within us as Apollo's lyre, in the form of concrete laws that are much stronger than natural laws. In us, the cosmos is playing *Apollo's lyre*. The human body is more than simply the aspect that biology recognizes; it is also a most wonderful musical instrument."[157] (emphasis added)

We may wonder why this inner music, which appears only in the etheric and astral bodies, has physical receptors similar to those in the inner ear. On the one hand, as we shall soon see, we also perceive only the etheric aspect of external music through the inner ear (see p. 87). On the other hand, the CSF-contacting neurons reflect the fact that the experience of inner music takes place in ordinary consciousness and does not require the development of higher, suprasensory perception ("Inspiration" in the anthroposophic sense). The experience of music in respiration is an experience in the etheric body, associated with neurosensory processes in the physical body. Thus music builds the bridge between the sensory world and the spiritual world. It is an imaginative experience of soul-spiritual (that is, inspirational) configurations, but it is granted to ordinary consciousness.

On this basis, we can understand why composers can be deaf (for example, Beethoven, Dvořák, and G. Fauré); whereas it is inconceivable that a painter could be blind. Music is not primarily an experience in the physical world. Composers take hold of it as described above, in the ebb and flow of creatively active chemical-etheric forces, which they breathe, and which are reflected in the physical body.

The Liver, on Listening to Symphonies

The limb character of hearing and the musical nature of metabolic chemistry, as described in Chapters 1 and 2 respectively, enter our willed, feeling consciousness through respiration. As the center of the metabolic system, more than any other organ, the liver is richly supplied with venous blood, which makes up 70 to 80 percent of the organ. The liver is attached to the diaphragm, so its fluids are directly affected by respiratory movement. Through respiration, the liver's chemical activity rises to consciousness (as sound ether) in the CSF. As a result, we can begin to understand why "there is an important difference between the liver of a musician and the liver of a non-musician, because the liver has a great deal to do with the musical images that resound within the human body." The liver is "the seat of everything active in the consequences of a beautiful melody, and the liver has a great deal to do with listening to a symphony. We must simply realize that the liver also includes an etheric aspect, which is primarily involved in these phenomena."[158] If we then consider how food and water connect the liver to influences from the person's environment, we can easily conclude that the chemical-etheric forces of geological features work into the liver. According to Rudolf Steiner's research, the specific geology of the Vienna Basin was what made Vienna "a gathering place for all…musicians." "You must have a sense of the intimate relationship between spiritual elements and the land. Consider the meaning of the fact that all of the geological features of Europe essentially come together in Vienna. Then consider that the relationships among substances are actually a musical scale—after all, relationships among atomic weights are the same as intervals, aren't they? (See pp. 31 ff.) If you consider all this, you will see that we are right on the mark when we say, based on cosmic circumstances, that the soul-spiritual milieu in Vienna is one where truly exceptional musical geniuses can feel at home."[159]

Lunar Forces in the Human Body and in Music

In conclusion, let's take another look at the anatomical structure of the ear. The sensory receptors in the ear have no contact with the airwaves produced by sound. They react only to the resonance the air vibrations create in the fluid of the inner ear. Let's compare this situation to how the sensory cells of the other sense organs work. The receptors of the senses of sight, smell, taste, and touch interact directly with stimuli from the outer world. Light lands on the ends of cones and rods in the retina; the molecules that produce smells and tastes interact with chemoreceptors; and pressure acts on touch receptors in the skin.

In order to interact in this same way with physical vibrations in the air, the auditory receptors would have to be located on the eardrum or in the outer auditory canal itself. This type of hearing actually does exist: in the sensilla of insects, where sensory hairs vibrate in lock step with the air vibrations of sounds. In vertebrates and humans, however, the ear's sensory activity is displaced inward, out of the air and into the water.

"The ear is actually the organ that deflects sound, which is originally active in air, into the interior of the body, separating it from the element of air so that the sound we then hear is active in the ether element. *The purpose of the ear, therefore, if I may put it like this, is to overcome air-based sound and to reflect the purely etheric experience of sound into the interior of the body.*"[160] On its path into the interior of the petrous bone, the meaning-bearing, ensouled, and living being of sound in the sound ether lays down its physical body of air in a death process similar to what we have seen in the production and the movement of CSF. In the inner ear and brain, the living soul and spirit being of sound dies and awakens to its etheric nature—in our consciousness. In hearing, we take part in the etheric, living activity of sounds, not in physical reality as we do in seeing, smelling, touching, and tasting. In hearing, therefore, we live in a relationship to the world of sound that applied to all the senses during the Old Moon epoch. In this sense, hearing is an atavism. When the fluid Moon stage was recapitulated during the Lemurian age of Earth evolution, "the human life-body as the receiver of sound guided the flow of air" in respiration. At that time, human respiration (and with it, the human body itself) were shaped according to the sounds the etheric body received.[161] This is the archetype of the musician, who "breathes" living experiences of sounds.

In this Lemurian age, the physical body was still much softer than it is today. Its organic forms were more flexible; they were dependent on the person's momentary soul-constitution to an extent that remains true today only with regard to our facial expressions, posture, and gestures. During respiration, the Lemurian human being reconfigured the entire fluid, flexible body, as is still the case today with venous blood and cerebrospinal fluid.

"[At that time] the soul still experienced the assimilation of air in a purely soul-spiritual way, as a pictorial process appearing in the form of ebbing and flowing images of sound that supplied the germinal nucleus with forms as it was differentiating. The soul felt surrounded by surging sounds on all sides and sensed how it shaped its body in accordance with these sound-forces."[162] That was how human beings in Lemurian times experienced themselves in breathing. Today, we recapitulate this phase during embryonic development. As we grow and mature, the formative forces of the spiritual music of the spheres, which have assumed physical form in our body, are released and appear in the artistic activity of creating and experiencing music. Music reveals the "secret laws of nature" in the human body. Without music, these laws would remain hidden for eternity.[163] Modern embryonic development and the Lemurian stage of Earth evolution are both metamorphoses of the Moon Epoch in the evolution of the Earth and humanity.

Human consciousness on the Old Moon was subject to a rhythmic alternation. During that stage of evolution, the Earth rotated slowly, so our human ancestors alternated between times of facing the Sun and times of facing away from it. This alternation was reflected in the circumstances of their life and in their consciousness. During Sun times, humans "became more devoted to the Sun and its phenomena than to themselves. At such times they experienced and absorbed, so to speak, the grandeur and glory of the cosmos as expressed in the existence of the Sun. In this case, the exalted beings who had their dwelling on the Sun were working...."[164] During this state, the physical-etheric body was regenerated. Self-awareness receded and imaginative *pictures* disappeared, to be replaced by *awareness of sounds*. During the period of turning away from the Sun, human beings were more concerned with themselves. The astral body became active, and the dreaming sound-consciousness of the Sun period was replaced by a more wakeful image consciousness. The physical-etheric body also gradually hardened during the stage of turning away from the Sun and toward Moon existence. This period is a state midway between being born and awakening as we now know it. "Similarly, the gradual dimming of consciousness as the Sun period approached resembles a state halfway between falling asleep and dying, because on the Moon, human consciousness of birth and dying was not yet what it is today. During a kind of Sun life, human beings gave themselves up to enjoying that life and were withdrawn from their own life for the duration, living more spiritually. It is possible to attempt only an approximate and figurative depiction of what the human being experienced during these times. He felt as if the causative forces of the cosmos were streaming into himself and pulsating through himself. He felt as if drunk with the cosmic harmonies he lived within. During these times, the astral body was as though freed from the physical body, and a part of the life body also withdrew from it. This formation consisting of astral and life body was like a marvelous, delicate musical instrument, and the mysteries of the universe resounded from its strings.... The sun beings were at work in these harmonies."[165] This "marvelous musical instrument" of the Old Moon stage was transformed into "Apollo's lyre" in the human beings of the Earth epoch of evolution. These secrets resounded into the Pythagorean Mysteries that Goethe imprinted on his immortal words about the spiritual rising of the Sun (*Faust II*, verses 4666-4668):

> Hear ye, hear ye the storm of Horae!
> Sounding as for spirit ears,
> The new day is born.

The genesis of "Apollo's lyre" explains why Rudolf Steiner, after describing it in Munich and Vienna as the respiratory oscillation of the cerebrospinal fluid that occurs when music is played or experienced, went on to embed it in the relationship of today's Earth to today's Sun. He described how breathing is incorporated into the displacement of the vernal equinox, and the role of the so-called Platonic world-year in musical experience.[166]

Respiration between Sun and Earth

The growth that arises from the earth in spring and summer, obeying the pull of the sun and stars, flows out into the surrounding atmosphere during St. John's Tide [from the Festival of John the Baptist on June 24, and the days following] in the scent of the flowers and the dispersal of the pollen: the Earth's soul is exhaling. With fructification in summer, inhalation begins. Fruits grow heavy and fall to the ground, sinking their seeds into the soil to rest during the long winter's night. The inhalation and exhalation of the Earth's soul and vegetation follows the course of the year, which is bound to the vernal point; that is, the point at which the celestial equator (the Earth's equator projected out into the cosmos) intersects the plane of the ecliptic, on which the zodiac lies. The vernal point, which is now located in the constellation Aries, moves through the entire zodiac once in every 25,920 years (this is a so-called Platonic world-year). In the course of one day, the human lungs take 25,920 breaths (18 x 60 x 24), infusing air into and out of the bloodstream.

Although these two phenomena are totally different from the physical perspective; and although the time-scales of the Earth's breathing in the world-year and the lung-breathing of a human day are very dissimilar, the same *number* is at work in both of these instances. Beyond all spatial, temporal, physical, and etheric differences, the same spiritual laws are at work in cosmic astral sensation, integrating the Earth's respiration and the respiration of the human lungs into a single whole. The laws of the number ether, in the form of the sound ether, are accessible to the astral body's musical perception.[167]

"I believe there is no abstract religious idea capable of arousing as much fervor as the awareness that the outer physical body is incorporated in this way into the macrocosm, into the very structure of the cosmos. Seers attempt to penetrate this phenomenon on the spiritual level. Within us, music emerges, rising up out of the body and into the soul. The soul's resounding, in harmony with the cosmos, is the unconscious element in artistic creativity. The entire cosmos resounds with us when we are truly artistically creative."[168]

The Experience of Music and Its Basis in Physiology 91

The anthroposophic mathematician Georg Glöckler has revealed the inner life of the number 25,920.[169] When he identified its divisors, starting with 24, the results were absolutely astonishing:

MAJOR SCALE ↓	24	x	1080	↑ MINOR SCALE	=	25.920
	27	x	960		=	25.920
	30	x	864		=	25.920
	32	x	810		=	25.920
	36	x	720		=	25.920
	40	x	648		=	25.920
	45	x	576		=	25.920
	48	x	540		=	25.920

And when these divisors are set in proportion to 24, the result is the proportions of the intervals of the major scale!

MAJOR ↓	24 : 24	=	1 : 1	Prime
	24 : 27	=	8 : 9	Second
	24 : 30	=	4 : 5	Major third
	24 : 32	=	3 : 4	Fourth
	24 : 36	=	2 : 3	Fifth
	24 : 40	=	3 : 5	Major sixth
	24 : 45	=	8 : 15	Major seventh
	24 : 48	=	1 : 2	Octave

When divided by 540, the series of larger divisors in the column on the right in the first chart, above, results in an interesting minor scale in which the major second is replaced by the half-tone (15:16) and the minor seventh is also different (9:16). The remaining intervals are the same:

MINOR ↓

540 : 540 =	1 : 1	Prime
540 : 576 =	15 : 16	Half-tone
540 : 648 =	5 : 6	Minor third
540 : 720 =	3 : 4	Fourth
540 : 810 =	2 : 3	Fifth
540 : 854 =	5 : 8	Minor sixth
540 : 960 =	9 : 16	Minor seventh
540 : 1080 =	1 : 2	Octave

When 360 is chosen as the divisor, the following proportions result.

Intervals of the minor scale:

360 : 360 =	1 : 1	Prime
360 : 405 =	8 : 9	Second
360 : 432 =	5 : 6	Minor third
360 : 480 =	3 : 4	Fourth
360 : 540 =	2 : 3	Fifth
360 : 576 =	5 : 8	Minor sixth
360 : 648 =	5 : 9	Minor seventh
360 : 720 =	1 : 2	Octave

We breathe within the inner framework of the Platonic world-year in the number ether. This framework underlies European music from Johann Sebastian Bach to Gustav Mahler in the form of a law that brings the human being and the cosmos into harmony. Here the "Jacob's ladder" of the astral body (see fig. 21, p. 43) reveals its cosmic and human reality! In the rhythm of our breathing, Apollo's lyre is tuned to the Platonic world-year.[170]

"The body is incorporated into the world in a fashion that corresponds to cosmic harmony. The soul, when it dwells in sensory perception and the activity of ordinary reason, is bound to the body with a strength that allows the body to transmit its cosmic harmony to the soul."[171] We experience this reality in great works of music.

Afterword

The Unique Status of Central Europe's "Classical" Music

Since classical music appears only in a very specific time and culture (that of Central Europe), some readers may question the justification of elevating the diatonic scale with its major and minor harmonies to a single valid system of music, so to speak. To answer this question, we will need to consider the position of classical music in the history of music, on the basis of Rudolf Steiner's spiritual science, as Hermann Pfrogner has done.[172]

Over the course of five to eight millennia, humanity's consciousness has shifted more fundamentally than is generally acknowledged today. Anthroposophic spiritual science looks back even further, to times before the first historical documents were handed down. At that time, physical bodies were more delicately constituted (see pp. 87 ff.), which is why there are no fossil remains of human beings from the early Atlantean age, for example. In Atlanteans, the soul-spiritual element was only loosely connected to the body. When they experienced music, they were immediately transported out of their bodies in ecstasy and entered a trance-like state. According to Rudolf Steiner, Atlantean music was experienced in scales built on the interval of the seventh. Earlier still, in Lemurian times, the experience of music was a spiritual participation in the cosmic music of the spheres, which sounded in scales based on ninths.[173]

As evolution progressed, the intervals experienced in musical scales became ever narrower. For instance, Chinese music is based essentially on the experience of the fifth and retreats into its original state, the *TAO* of the spiritual harmonies of the cosmos.[174] As suprasensory consciousness was lost, the soul entered ever more deeply into the physical body. *Musica mundana* (the music of the spheres) was transformed into *musica humana*. In the fifth, *musica mundana* lives in the breathing process, that is, on the boundary between the physical body and the atmosphere.[175]

After the beginning of the Christian era, the people of Europe experienced music in increasingly narrow intervals, and as increasingly connected to their physical bodies. In Gregorian music, the underlying fifth is still prominent and sets the mood; at that time the third was still experienced as dissonant. The major third could be experienced as consonant only once the soul became able to unite with the "inward extension of the experience of breathing"; that is, with the blood.[176] The experience of major and minor as we know it since the Renaissance and Baroque periods, and in the music of Haydn or Mozart, for example, derives from a constitution of the soul (that is, of the astral body) that identifies fully with the blood in the rhythmic system of the heart and lungs. (Physiologically, the proportion of blood to air corresponds to the so-called ventilation/perfusion ratio of 4:5.)[177]

With Beethoven, the incarnation of music in Europe took a step further. In the passages most characteristic of his work, Beethoven identified with the physical body's metabolic-limb forces. Goethe, who chose to remain entirely in the central, rhythmic system in his appreciation of art, had no sympathy for this aspect of Beethoven's music. "It makes you think the house is falling down," Goethe said when Mendelssohn, at the age of twelve, played Beethoven's *Fifth Symphony* for him on the piano. Clearly, Goethe already sensed that this music was making inroads into the physical nature of the human being.[178]

Josef Matthias Hauer, a contemporary of Arnold Schönberg, who also developed twelve-tone music in Vienna, experienced this development of music as problematic or even ethically questionable. For Hauer, the route music followed from the cosmos of Chinese Taoism through Greece to "Viennese classicism" was nothing other than music's "fall from grace"—the human soul's increasing identification with the material, physical body. It was the path that connected the cosmic being of *melos* with physical "noise," along with the progression from twenty, then fifty, to a hundred musicians deemed necessary to allow a "symphony" to resound. For Hauer, this was the path "from *melos* to kettledrums," as he put it in one of his program notes.[179]

In Romanticism, the sound of the instruments became ever more important, achieving its greatest refinement in the instrumentation of

Wagner, Debussy, or Ravel, for example. Orchestras continued to increase in size until the death of this trend in European music. Gustav Mahler saw it to its grave, in part with sarcastic humor, but also with deep sensitivity that reached across the threshold.

Vienna: The Earthly Home of Classical Music

Why did this phase in the development of music happen in Vienna, of all places on earth? It has always been a puzzle why Hydn (from Eisenstadt), Mozart (from Salzburg), Beethoven (from Bonn), Brahms (from Hamburg), Bruckner (from Linz) and Mahler (from Böhmen all settled in Vienna; only Schubert was Viennese by birth. According to Rudolf Steiner, the geology of the Viennese basin facilitated the creative process for musicians working there (see p. 86). Think, for example, of the radiant effects of the earth's chemical ether and of the chemistry of the groundwater in the metabolic-limb system in the sense described in Chapters 1 and 2. In Vienna, the being of Music united with the earthly forces of the physical body. Dying into the body, it became the chord:

Example 11: Ludwig van Beethoven, Coriolan Overture, Opus 62.

(See, A. Husemann, *Der musikalische Bau des Menschen*, 4th edition 2003 p. 159-163. In English as *The Harmony of the Human Body*.)

It fell to Beethoven to take the step from living to mechanical tone production, from *musica instrumentalis* to the piano. What this all means, however, is that Viennese classicism aligned music with the physical body. With its major-minor tonality, this music, more than any other style, opens the door to a musical physiology, to the mysteries of the physical body. Investigating the musical systems of other cultures and times would provide access to the more spiritual layers of human nature. It is certainly no coincidence, however, that popularized forms of this music have been so successfully adapted to technology and have fed a worldwide marketing industry. Rudolf Steiner provided multifaceted suggestions for the further development of music in our time. Michael Kurtz is now at work on a book on that subject (*Rudolf Steiner und die Kunst der Musik* [Rudolf Steiner and the art of music], not yet published.)

Glossary:

Å	angstrom; a unit of length equal to 10^{-7} mm
actin, myosin	the two types of protein filaments that interact in muscle movement; see also "filament"
Biofeedback	the use of monitors to provide information about biological processes in sense-perceptible form (e.g., heart-rate monitor)
Bradykinesia	slowed ability to start and continue movements
burning feet syndrome	a medical condition that causes burning sensations in the feet
Calvarium	the upper domelike portion of the skull
Cerebellum	the part of the brain at the back of the skull, responsible for coordinating movements
cerebrospinal fluid	CSF; the clear fluid filling the space between the arachnoid membrane and the pia mater (membranes surrounding the brain)
choroid plexus	a network of capillaries located at the base of a ventricle, where CSF forms in a filtration process similar to that of the kidneys

CSF-contacting neurons	Different types of sensory cells located in the walls of the brain ventricles and in the arachnoidal space. They convert the composition and movement of CSF into nerve impulses.
CSF system	the totality of space within and around the brain and spinal cord in which CSF circulates and is subject to respiration-dependent movement.
cuticular plate	a laminar structure in the in the hair cells of the organ of Corti from which their stereocilia protrude
cytoskeleton	the fibrous supporting structure in cell walls. The cytoskeletons of the outer hair cells contain high-speed motor proteins.
D^{30}	In the customary homeopathic abbreviations for decimal potencies, a D1 solution is produced when 1 part of the initial substance is mixed with 9 parts of a diluent (e.g., water). Repeating this process 30 times produces a D30 solution. At each potency level, the new mixture is shaken rhythmically for three minutes.
depolarization	the discharge of electrical tension that builds up in a nerve due to unequal distribution of ions; equivalent to "excitation" of a neuron.
discrimination	here, the ability to distinguish differences in the height of pitch

dura mater	the tough fibrous membrane covering the brain and the spinal cord and lining the inner surface of the skull
dysesthesia	impaired sensation in the skin
ectoderm	the outermost of the three primary germ cell layers that make up the early embryo
embryonic period	the time from conception to the end of the third month of gestation; see also "fetal period."
endoderm	the innermost layer of cells in the early embryo; see also "ectoderm"
endolymph	the fluid in the organ of Corti in the inner ear
endolymphatic duct	the channel connecting the inner ear with the endolymphatic sac under the dura mater, allowing changes in pressure in the endolymph to be transmitted to the cerebrospinal fluid.
encephalitis	inflammation of the brain
ependyma	the membrane lining the ventricles of the brain
extradural	located outside the dura mater of the spinal cord
fetal period	the time from the fourth month of gestation until birth; see also "embryonic period"
filament	a thread-like proteinaceous structure

hair cells	In a cross-section of the organ of Corti, there are three rows of outer and one row of inner *hair cells*, the ear's sensory receptors
hypertension	(arterial) high blood pressure
Imagination	In Rudolf Steiner's spiritual science, perception of etheric formative forces through body-free imagination
imaginative	Living sensory realities can be thought of "imaginatively," as Goethe did in his plant morphology and color theory. The life process is grasped as a living idea, not as an Imagination.
Imponderables	An archaic term for heat, light, and sound, because they have no weight. In spiritual science, the term also includes etheric, astral, and spiritual forces in the narrower sense.
Inner ear lymph	the endolymph of the cochlea; see also perilymph and endolymph.
Interval proportions	In terms of physics, an interval comes about as a whole-number ratio of two tone frequencies. In an octave, the ratio of the frequencies is 1:2; in a fifth, 2:3, in a fourth, 3:4, etc., as illustrated by the division ratios of a vibrating string.
Labyrinth	The organs of equilibrium and hearing are housed in a common space filled with perilymph. This space is called the "labyrinth" due to its complex shape.

Loop diuretics	diuretics that act on the loop of Henle in the nephrons
Marsyas	The demigod Marsyas was defeated by Apollo in a musical competition, whereupon Apollo flayed him alive.
Meningitis	inflammation of the meninges, the membranous covering of the brain and spinal cord
mesoderm	in the early embryo, the middle germ layer from which blood vessels and muscles develop; see also "endoderm" and "ectoderm."
nerve potential	see "potential"
operculum	In amphibians, a small cartilaginous plate that rests on the oral window and transmits sound conducted from the ground to the forelimbs and shoulders to the eardrum via the opercular muscle.
ossification	bone mineralization, the process of bone tissue formation by cells called osteoblasts
perilymph	the fluid in which the cochlea of the inner ear floats. Perilymph, which is chemically similar to CSF, fills the gap between the bony cochlear canal and the organ of Corti. See also "perilymphatic duct."

perilymphatic duct	the connection between the perilymphatic space of the cochlea and the external cerebral fluid, also called the "cochlear aqueduct." Its width varies greatly in individuals.
pineal gland	the endocrine gland in the brain, producing melatonin
primitive gut	the primordium of the intestinal tract in the embryo. Invaginations of the primitive gut develop into the lungs and the liver.
restless leg syndrome	pathological, uncomfortable sensations in the legs while sitting or lying
retina	the light-sensitive membrane lining the inner eyeball
rigidity	stiffness of a limb
Scala vestibuli	The acoustic pressure wave passes through the scala tympani (the upper half of the cochlear canal) to the apex of the cochlea but then it turns back and exits the inner ear via the *scala vestibuli*, through the round window.
semicircular canals	the three tubes in the inner ear that provide sensory input for experiencing movement in space. See also "labyrinth."
sensory nerves	nerves that conduct impulses from sensory organs toward the brain

stapedius muscle	the tiny muscle that pulls the head of the anvil away from the oval window of the cochlea as a protective reflex, e.g., in case of sudden loud noise
stirrup	the innermost of the three auditory ossicles (hammer, anvil, and stirrup); also called "stapes"
tensor tympani muscle	a tiny muscle that inserts into the hammer and regulates the tension on the eardrum
tympanic cavity	the space between the eardrum and the inner ear, bridged by the auditory ossicles; also called "middle ear."
Type	used here in the Goethean and Steinerian sense
varicosis, varicose veins	enlarged, tortuous veins, especially on the legs
venous ulcer	wound resulting from venous insufficiency
ventricle	here: one of four systems of fluid-filled spaces in the brain

References

1. A. Brendel, *Musik beim Wort genommen*, Munich: Piper 1992, p. 26.
2. See Peter Selg, *Vom Logos der menschlichen Physis. Die Entfaltung einer anthroposophischen Humanphysiologie im Werk Rudolf Steiners*, Dornach 2000, pp. 309, 469, 515ff.
3. R. Steiner, *Die Philosophie der Freiheit*, GA 4, Dornach: Rudolf Steiner Verlag 1962, p. 270 Also in English as *Intuitive Thinking as Spiritual Activity*, Anthroposophic Press.
4. M. Bindelli, "Arbeitswoche für Menschenwissenschaft durch Kunst" [Conference on the science of the human being through art], Stuttgart. 1995. Topic: The Hearing Human. See A. Husemann (ed.), *Menschenwissenschaft durch Kunst. Die plastisch-musikalisch-sprachliche Menschenkunde*, Stuttgart: Verlag Freies Geistesleben 2007, pp. 223, 226.
5. E. Kolb (ed.), *Lehrbuch der Physiologie der Haustiere*. 5th edition, Stuttgart: Gustav Fischer 1989, p. 949.
6. F. Hollwich, *Augenheilkunde*. 8th edition, Stuttgart: Thieme 1976, p. 3.
7. E. Gaupp, Beiträge zur Kenntnis des Unterkiefers der Wirbeltiere (1911). See D. Starck, *Vergleichende Anatomie der Wirbeltiere*, vol. 2, Berlin: Verlag Julius Springer 1979, pp. 336ff.
8. G. Husemann, "Arbeitswoche für Menschenwissenschaft durch Kunst," Stuttgart 1982. Topic: The ear and the larynx. See note 4.
9. G. Heldmaier, G. Neuweiler, *Vergleichende Tierphysiologie*, vol. 1. Berlin, Heidelberg, New York: Julius Springer 2003, pp. 250ff.
10. A. Tumarkin, Evolution of the Auditory Conducting Apparatus in Terrestrial Vertebrates, in: A.V. S. de Reuck, J., Kringht (eds.), *Hearing Mechanisms in Vertebrates. A Ciba Foundation Symposium*, London: J & A Churchill Ltd. 1968, pp. 18-37; A. Portmann, *Einführung in die vergleichende Morphologie der Wirbeltiere*, 5th edition, Basel, Stuttgart: Schwabe 1976, p. 171, fig. 149.
11. Heldmaier, Neuweiler, op. cit., pp. 248ff.
12. K. M. Douglas, D. K. Bilkey, Amusia is associated with deficits in spatial processing, in: *Nature Neuroscience*, Vol. 10, no. 7, July 2007, pp. 915-921.

13. H. P. Zenner, Physiologie und biochemische Grundlagen des normalen und des gestörten Gehörs, in: H. N. Naumann, J. Helms, C. Herberhold (eds.), *Oto-Rhino-Laryngologie in Klinik und Praxis. Bd. I: Ohr*, Stuttgart: Thieme 1994, pp. 81-259.

14. Ibid. According to a personal communication with the author in November 2005, research to date had confirmed the findings referenced here on the physiology of the inner ear.

15. R. F. Schmidt, G. Thews, *Lehrbuch der Physiologie*, 29th ed., Berlin, Heidelberg, New York: Julius Springer 2005.

16. Ibid.

17. See note 13.

18. See note 13.

19. Goethe on his theory of sound: *Naturwissenschaftliche Schriften*, R. Steiner (ed.), vol. 5, Dornach: Rudolf Steiner Verlag 1982, p. 598.

20. See note 13.

21. R. Steiner, *Das Wesen des Musikalischen und das Tonerlebnis im Menschen*, GA 283, 5th edition, Dornach: Rudolf Steiner Verlag 1989, Lecture 1 (March 7, 1923). Also in English as *The Inner Nature of Music and the Experience of Tone*.

22. See note 8.

23. The fact that music and speech resound in space is irrelevant to their content. A melody or a spoken sentence remains the same whether I hear them coming from the right or the left, from in front of me or behind me. Space is essential to limb movement, but to the movement of music and speech, it is simply the medium in which they appear. Their content and their characteristic process flow in *time*.

24. S. Hahnemann, *Organon der Heilkunst*, 6th edition, Coethen 1865, § 20, p. 112.

25. The process of "turning inside out," taken from projective geometry, leads our thinking to into the reality of etheric space. See G. Adams, O. Whicher, *Die Pflanze in Raum und Gegenraum. Elemente einer neuen Morphologie*, Stuttgart: Verlag Freies Geistesleben 1979.

26. R. Steiner, *Geisteswissenschaft und Medizin*, GA 312, Dornach: Rudolf Steiner Verlag 1976, Lecture 2 (March 22, 1920). Also in English as *Introducing Anthroposophical Medicine*, SteinerBooks,

27. T. McKeen, Die Anatomie des Ohres und das Hören, in: *Wesen und Gestalt des Menschen*, Stuttgart: Verlag Freies Geistesleben 1996, pp. 204-223.

This essay's essential content is also based on Gisbert Husemann's lectures (see note 8).

28. At ultrasound frequencies, larynx function in bats and dolphins metamorphoses into a *sensory* – i.e., head – function (sonar). This is one example of Rudolf Steiner's statement that when an entity – in this case, sound frequency – is augmented, its effects never retain the same quality into infinity but are *reversed* at a certain point, so that the opposite quality appears as augmentation continues (Steiner, op. cit., GA 312).

29. W. S. Condon, L. W. Sander, Synchrony Demonstrated Between Movements of the Neonate and Adult Speech, in: *Child Development*, Vol. 45, 1974. As quoted in E-M.Kranich, *Der innere Mensch und sein Leib*, Stuttgart: Verlag Freies Geisteslebon 2003.

30. Kranich, ibid. R. Patzlaff, *Der gefrorene Blick*, Stuttgart: Verlag Freies Geisteslebon 2004.

31. I. Molnar-Szakacs, K. Overy, Music and mirror neurons: from motion to 'e'motion, in: *Social, Cognitive and Affective Neuroscience* 2006 1(3):235-241.

32. Pacchetti et al., Active music therapy in Parkinson's disease: an integrative method for motor and emotional rehabilitation, in: *Psychosom. Med.* 2000/62 (3), pp. 386-393.

33. O. Sacks, *Musicophilia: Tales of Music and the Brain*, New York: Knopf 2007, pp. 248-258.

34. T. Wehner, S. Vogt, M. Stadler, P. Schwab, P. Kruse, Intra- and Interpersonal Biosignal Processing: Further Developments of Common EMG Biofeedback Procedures, in: *Journal of Psychophysiology* I, 1987, pp. 135-148.

35. Just as the ear's structure is analogous to the metabolic-limb system, the eye is anatomically analogous to the brain, and in embryological development it originates in the diencephalon. The retina is an extension of the brain, which is why we can apply maximum distance and critical ability to visual impressions. See Gisbert Husemann's study "Gehirn und Auge" in: *Sinnesleben, Seelenwesen und Krankheitsbild*, Stuttgart: Verlag Freies Geisteslebon 1998, pp. 18-27.

36. O. Sacks, *A Leg to Stand On*, New York: Touchstone 1984, p. 116

37. Starck, op. cit., vol. 3, p. 657.

38. C. Breme, *Plastisch erarbeitete Embryologie*, 3rd edition, Leymen: AAP Verlag 2008, p. 28. Illustration used with permission of the author.

39. R. Steiner, Meditativ erarbeitete Menschenkunde, in: *Erziehung und Unterricht aus Menschenerkenntnis*, 4th edition, GA 302a, Dornach: Rudolf Steiner Verlag 1993, Lecture of September 21, 1920. Also available in English as *Balance in Teaching*.

40. V. Zuckerkandl, *Die Wirklichkeit der Musik*. Zurich: Rhein-Verlag 1963.
41. A. Schopenhauer, *Die Welt als Wille und Vorstellung*. *Drittes Buch. Die Platonische Idee. Das Object der Kunst*, in: *Werke*, edited by R. Steiner, (Stuttgart n.d.1894), vol. 5/6, p. 110 (emphasis added).
42. Steiner, *Das Wesen des Musikalischen und das Tonerlebnis im Menschen*, loc. cit., see note 21. Also in English as *The Inner Nature of Music and the Experience of Tone*.
43. R. Steiner, *Der irdische und der kosmische Mensch*, GA 133, Dornach: Rudolf Steiner Verlag 1989, lecture of June 18, 1912. Also in English as *The Earthly and the Cosmic Human Being*.
44. As a Buddhist, Schopenhauer believed in reincarnation.
45. Brendel, op. cit. (note 1), p. 26.
46. R. Steiner, *Geistige Zusammenhänge in der Gestaltung des menschlichen Organismus*, GA 218, 3rd edition, Dornach: Rudolf Steiner Verlag 1992, lecture of December 9, 1922. Also in English as *Waldorf Education and Anthroposophy*, vol.2.
47. J. Staubesand, Kopf und Hals, in: *Benninghoff. Anatomie*, vol. 1, 14th edition, Munich, Vienna, Baltimore: Urban und Scwarzenberg Verlag 1984, p. 502 (emphasis added).
48. A dissertation investigating this issue in greater detail on the basis of existing data would be well worthwhile. A related phenomenon is the specifically human shortage of space for permanent teeth at the front of the jaw. The jaw orthopedist Stöckli expresses surprise at the absence of exact genetic correlation between the dimension of the jawbone and the sum of the widths of the teeth. Here, too, Rudolf Steiner points to the I-organization, which is active in the second dentition but not in the first. See A. Husemann, Der Zahnwechsel des Kindes, Stuttgart: Verlag Freies Geistesleben 1996, pp. 112ff.
49. R. Steiner, *Menschengeist und Tiergeist*, GA 60, 2nd edition, Dornach: Rudolf Steiner Verlag 1983, lecture of November 17, 1910. Also in English as *Education of the Child*.
50. R. Steiner, *Geistige Zusammenhänge in der Gestaltung des menschlichen Organismus*, loc. cit. (see note 46).
51. R. Steiner, *Anthroposophische Leitsätze*, GA 26, Dornach: Rudolf Steiner Verlag 1972, no. 36, p. 29. Also in English as *Anthroposophical Leading Thoughts*, Rudolf Steiner Press.
52. R. Steiner, *Der Mensch als Zusammenklang des schaffenden, bildenden und gestaltenden Weltenwortes*, GA 230, 7th edition, Dornach: Rudolf Steiner

Verlag 1993, lecture of October 21, 1923. Also in English as *Harmony of the Creative Word*, Rudolf Steiner Press.

53. R. Steiner, *Über Gesundheit und Krankheit*, GA 348, Dornach: Rudolf Steiner Verlag 1959, lecture of November 29, 1922; and Steiner, conference of December 4, 1922 (GA 300/2, Dornach 1975). On July 2, 1921 (GA 205), Steiner described the brain as the re-embodiment of the liver from the previous earthly life. From the overall context of Steiner's descriptions of invagination processes in reincarnation that transform the lower organs into parts of the head, it is evident that the cochlea of the inner ear is the re-embodiment of the intestine.

54. There are also natural scientific correlates of this metamorphosis. The microvilli of intestinal cells are bathed in intestinal fluid just as the sensory hairs of auditory hair cells are bathed in endolymph. Just as sound is perceived in the ear, so too food is "tasted" unconsciously by receptors in the glycocalyx of the villi. "Antibodies against chicken fibrin show an affinity for auditory sterocilia in all the mammals and non-mammals studied" (Zenner, loc. cit., note 13). In 1985, W. Schad had already pointed out that the spiral forms of the intestines and ear develop at the same time. (W. Schad, Die Ohrorganisation, in: *Goetheanistische Naturwissenschaft* Vol. 4: *Anthropologie*, W. Schad (ed.), Stuttgart: Verlag Freies Geistesleben 1985.

55. In the interim, the effect of gravity on the embryonic development of plants and animals has been thoroughly investigated. For example, see Wolpert, *Entwicklungsbiologie*, Heidelberg/Berlin: Spektrum Akademischer Verlag 1999, p. 80 and pp. 236ff.

56. R. Steiner, GA 218, loc. cit. (see note 46).

57. R. Steiner, *Die Kunst des Erziehens aus dem Erfassen der Menschenwesenheit*, GA 311, Dornach: Rudolf Steiner Verlag 1989, pp. 101ff. Also in English as *The Kingdom of Childhood*, SteinerBooks.

58. R. Steiner, *An Outline of Esoteric Science*, CW13, Hudson: Anthroposophic Press 1997, pp. 135ff. See also M. Basfeld, *Wie Denken uns zu Menschen macht*, Stuttgart: Verlag Freies Geistesleben 2011.

59. Steiner, ibid., p. 146.

60. G. Beeli, M. Esslen, L. Jäncke, When coloured sounds taste sweet, in: *Nature* 2005: p. 38.

61. G. Husemann, *Erdengebärde und Menschengestalt*, Stuttgart: Verlag Freies Geistesleben 1962, p. 190.

62. I. Wegman, R. Steiner, *Grundlegendes für eine Erweiterung der Heilkunst nach geisteswissenschaftlichen Erkenntnissen*, GA 27, Dornach: Rudolf Steiner

Verlag 1991, ch. 17. Also in English as *Extending Practical Medicine*, Rudolf Steiner Press.

63. See also O. Wolff, *Grundlagen einer geisteswissenschaftlich erweiterten Biochemie*, Stuttgart: Verlag Freies Geistesleben 1998, pp. 308-317.

64. Newlands noted that if the elements are arranged according to increasing atomic weight, every eighth element displays similar chemical and physical properties. He called this phenomenon the "law of octaves." Since the inert gases were unknown at that time, a sequence beginning with carbon as the "tonic" then continued through N(2), O(93), F(4), Na(5), Ca(6), and Al(7) before arriving at silicon, which is the "octave" of carbon and has similar properties. The structural function that carbon serves in organic chemistry is served by silicon in the mineral world. Thus any element could be taken as the starting point for an eight-step period. Factoring in the noble gases, as in electron pair theory, results in a different perspective on musical thinking, which addresses the dynamic relationship among elements.

65. See J. Mezger, *Gesichtete Homöopathische Arzneimittellehre*, vol. 1, 3rd edition, 1964.

66. See J. Mezger, *Gesichtete Homöopathische Arzneimittellehre*, vol. 1, 3rd edition, 1964.

67. L. Trueb, *Die chemischen Elemente. Ein Streifzug durch das Periodensystem.* Stuttgart 1996.

68. R. Steiner, *Geisteswissenschaftliche Impulse zur Entwickelung der Physik*, vol. II, GA 321, 4th edition, Dornach: Rudolf Steiner Verlag 2000, lecture of March 14, 1920. Also in English as *The Warmth Course*, Mercury Press.

69. W. Amelung, G. Hildebrandt, *Balneologie und medizinische Klimatologie*, vol. 1, Berlin, Heidelberg, New York, Tokyo: Julius Springer 1985, pp. 30-108. B. Roßlenbroich, *Die rhythmische Organisation des Menschen*, Stuttgart: Verlag Freies Geistesleben 1994.

70. R. Steiner, Blut ist ein ganz besonderer Saft, lecture of October 25, 1906, in: *Die Erkenntnis des Übersinnlichen in unserer Zeit und deren Bedeutung für das heutige Leben*, GA 55, 2nd edition, Dornach, Rudolf Steiner Verlag 1983. Also in English as "Blood is a Very Special Juice," in *Supersensible Knowledge*.

71. R. Steiner, Die Ätherisation des Blutes, lecture of October 1, 1911, in: *Das esoterische Christentum und die geistig Führung der Menschheit*, GA 130, 4th revised edition, Dornach: Rudolf Steiner Verlag 1995. Also in English as "The Etherization of the Blood," in *Esoteric Christianity and the Mission of Christian Rosenkreutz*.

72. R. Steiner, *An Outline of Esoteric Science*, loc. cit. (note 58).

73. In *An Outline of Esoteric Science*, Rudolf Steiner deals with the difficulty of conceiving of heat without matter; see also the work of the physicist M. Basfeld, *Wärme: Ur-Materie und Ich-Leib*, Stuttgart: Verlag Freies Geistesleben 1998.

74. See D. Bosse, *Die gemeinsame Evolution von Erde und Mensch*. Stuttgart: Verlag Freies Geistesleben 2002.

75. Here Goethe, in spite of his typical reserve, unambiguously articulates the fundamental alchemical thought he had accepted forty years earlier during his studies with Susanna von Klettenberg. According to the original Rosicrucian idea, every chemical interaction is the outer manifestation of a soul-spiritual process inside the human being, "and all the more so, since everywhere there is only one Nature." See also H. Beckh, *Alchymie. Vom Geheimnis der Stoffeswelt*, Dornach 1984, pp. 38ff; G. Husemann, Goethes Verhältniss zum Zinn; Goethes Verhältniss zur Geologie, in: *Sinnesleben, Seelenwesen und Krankheitsbild*, loc. cit. (note 35), pp. 261ff and 278ff; D. Bosse, *Goethes Initiation und die Ursphäre der Erde*, Studien und Versuche, Stuttgart: Verlag Freies Geistesleben 1995.

76. R. Steiner, *Perspektiven der Menschheitsentwickelung*, GA 204, Dornach, Rudolf Steiner Verlag 1979, lecture of April 23, 1921. Also in English as *Materialism and the Task of Anthroposophy*, Rudolf Steiner Press.

77. A. Blickle, E. Bindel, Das periodische System der Elemente als Schöpfungsurkunde, in: *Die Drei*, April/June 1948.

78. R. Steiner, *Kunst im Lichte der Mysterienweisheit*, GA 275, 3rd edition, Dornach, Rudolf Steiner Verlag 1990, lecture of December 29, 1914. Also in English as *Art as Seen in the Light of Mystery Wisdom*, Rudolf Steiner Press. See also the Afterword to this book, p. 000.

79. R. Steiner, *Eurythmie als sichtbarer Gesang*, GA 278, 3rd edition, Dornach: Rudolf Steiner Verlag 1990, lecture of February 20, 1924. *Eurythmy as Visible Song*, [PUBL??]

80. Ibid.

81. R. Steiner, *Meditative Betrachtungen und Anleitungen zur Vertiefung der Heilkunst*, GA 316, 3rd edition, Dornach: Rudolf Steiner Verlag 1987, p. 167. Also in English as *Course for Young Doctors*, Mercury Press.

82. Johann Heinrich Deinhardt considered the possibility that the soul prepares the body for the next life already in this lifetime and then carries it over the threshold of death. See R. Steiner, *Vom Menschenrätsel*, GA 20, 4th edition, Dornach: Rudolf Steiner Verlag 1957, pp. 64 and 167; in English as *The Riddle of the Human Being*; and R. Steiner, *Geistige Zusammenhänge in der Gestaltung des menschlichen Organismus*, op. cit (see note 46),

lecture of November 4, 1922: "Die verborgenen Seiten des Daseins und der Christus-Impuls."

83. R. Steiner, *Das Wesen des Musikalischen und das Tonerlebnis im Menschen*, GA 283, 2nd edition, Dornach: Rudolf Steiner Verlag 1975, lecture of March 7, 1923. Also in English as *The Inner Nature of Music and the Experience of Tone*.

84. The mysterious figure of Mignon in Goethe's *Wilhelm Meister* has been the subject of several different anthroposophically oriented interpretations. To me, it seems most likely that this figure points to the developing human being, since it also appears in Rudolf Steiner's image as a child striving toward the present from the future: "Let me appear to be until I become."

85. R. Steiner, *Geisteswissenschaft und Medizin*, op. cit. (see note 26), lecture of April 6, 1920. In English as *Introducing Anthroposophical Medicine*, CW 312, SteinerBooks.

86. S. Russell, J. Ickovics, R. Yaffee: Exploring Potential Pathways between Parity and Tooth Loss among American Women. *American Journal of Public Health*, vol. 98, no. 7: 1263-1270, July 2008.

87. R. Steiner, *Geisteswissenschaft und Medizin*, op. cit. (see note 26), p. 190. Also in English as *Introducing Anthroposophical Medicine*.

88. R. Steiner, *Heilpädagogischer Kurs*, GA 317, 8th edition 1995, Dornach: Rudolf Steiner Verlag, lecture of July 5, 1924. *Education for Special Needs*, Rudolf Steiner Press.

89. G. Sachse, G. Siebert, Zur Mundgesundheit, zum Ernährungsverhalten und zur sozioökonomischen Situation von Jugendlichen. *Deutsche zahnärztliche Zeitschrift* 41 (1986), 191-194.

90. Rudolf Steiner, *Geisteswissenschaft und Medizin*, op. cit. (note 26), p. 190. Also in English as *Introducing Anthroposophical Medicine*. Readers who wish to examine the methods connected with this presentation in more detail should refer to Rudolf Steiner's lectures on the principles of physiology (january 3 and 6, 1923) in *The Origin of Natural Science*, (CW 316).

91. R. Steiner, *Geisteswissenschaftlich Gesichtspunkte zur Therapie*, GA 313, Dornach: Rudolf Steiner Verlag 1963, lecture 4. Also in English as *Spiritual-Scientific Aspects of Therapy*; H. Koepke, *Das neunte Lebensjahr. Seine Bedeutung in der Entwicklung des Kindes*, Dornach: Verlag am Goetheanum 1993.

92. R. Steiner, *Eurythmie als sichtbarer Gesang*, op. cit.(note 79), lecture of February 20, 1924. Also in English as *Eurythmy as Visible Song*.

93. Mezger, op. cit. (note 65).

94. A. Husemann, *The Harmony of the Human Body*, op. cit. (note 66).

95. Many thanks to Beatrice Hallqvist, MD (academic year 2003/2004) for her contribution to this section. After learning about the musical fluorine process, she had the intuition to apply fluorine therapeutically in this instance, and she helped write the text of the case history. These descriptions of the fluorine process in human beings are by no means complete (the connection to the magnesium process, in particular, is still missing). The author's concern here was simply to present the aspect of the fluorine process that relates to limb movement in the sense of Chapter 1.

96. Johann Gottfried Herder, Viertes kritisches Wäldchen, Zweites Stück. *Werke*, vol. II. Munich: Hanser Verlag 1987, p. 153.

97. J. Kaiser emphasizes that this ability to cause the entire piece to appear was characteristic of Backhaus. J. Kaiser, *Grosse Pianisten unserer Zeit*, Gütersloh n.d., pp 62-65.

98. J. Uhde, R. Wieland, *Denken und Spielen. Studien zu einer Theorie der musikalischen Darstellung*, 3rd edition, Kassel 1990, pp. 176ff.

99. R. Steiner, *Anthroposophy, Psychosophie, Pneumotosophie*, GA 115, Dornach: Rudolf Steiner Verlag 1965, lecture of Novembr 4, 1910. Also in English as A Psychology of Body, Soul, & Spirit, SteinerBooks; C. Hueck, Zeitenkreuz und Geistesgeschichte. *Das Goetheanum*, no. 28 (2010), pp. 10-11.

100. E Kurth, *Musikpsychologie*,1931, reprinted 1969, Georg Olms Verlag, Hildesheim, New York: V. Zuckerkandl, op. cit. (note 40); J. Uhde, R. Wieland, op. cit. (note 98).

101. T. W. Adorno, *Aufzeichnungen zu einer Theorie der musikalischen Reproduktion*, part 2, p. 93, as quoted in Uhde, op. cit. (note 98), p. 232.

102. R. Steiner, *Mein Lebensgang*, GA 28, Dornach: Rudolf Steiner Verlag 2000, ch. 4. In English as *Autobiography*, SteinerBooks. See also M. Kurtz, Rudolf Steiner und Anton Bruckner, Part 1, in: *Stil*, Easter 2010/2011, vol. 32, issue 1. H. Kern pointed succinctly to the musical aspect of Goethe's thinking and thus to the methodological foundation for a science of music: Goethe's methods "put the researcher in a position to think and observe the nature of understanding in movement. In particular, 'contemplative judgment' as the organ for comprehending life is an eminently musical element." H. Kern: Goetheanismus und Musik, in: *Erziehungskunst* 7/8 (2009), p. 807.

103. See D. Ebert, Atmung beim Klavierspielen. *Therapiewoche* 15 (1996), pp. 800-805; D. S. Ellis, Effects of Music on Respiration and Heart-Rate. *Amer. J. Psychol.* 65 (1952), pp. 39-47.

104. H. H. Wängler, Höratmung Sprechatmung, in: *Zeitschrift für Phoniatrie* 1959 and 1973, as quoted in O. von Essen, *Allgemeine und angewandte Phonetik*, Berlin: Akademie Verlag, n.d., p. 14.

105. G. A. Roemer, Atmung und musikalisches Erleben (1925), in: L. Heyer-Grote (ed.), *Atemschulung als Element der Psychotherapie* (WdF v. 65) 1970, pp. 48-53.

106. E. Pöppel, *Wie kommt die Zeit in den Kopf?* http://www.ifa.de/pub/kulturaustausch/archiv/zfk-1998/zeit/poeppel

107. E. Pöppel, *Eine zu große Herausforderung? Einige Fragen über die Zeit. Forschung und Lehre*; http://www.forschung-und-lehre.de/archiv/12-99/poeppel.htm

108. E. Pöppel, *Wie kommt die Zeit in den Kopf?* op. cit. (note 106).

109. See also E. Pöppel, *Der Rahmen*, Munich: Hanser 2005, p. 302.

110. Remarkably, Pöppel also comes to the conclusion that space and time, rather than being *a priori* givens for human beings (as Kant believed), are *integrative accomplishments of thinking*, as Rudolf Steiner also suggested in Der Goethesche Raumbegriff, in: *Einleitungen zu Goethes Naturwissenschaftlichen Schriften*, Stuttgart 1962, pp. 207ff.

111. Pöppel, *Der Rahmen*, loc. cit. (note 109).

112. R. Steiner, *Das Rätsel des Menschen. Die geistigen Hintergründe der menschlichen Geschichte*, GA 170, 3rd edition, Dornach: Rudolf Steiner Verlag 1992, lecture of August 15, 1916. Also in English as *The Riddle of Humanity*, Rudolf Steiner Verlag; *Von Seelenrätseln* (1917), GA 21, 4th edition, Dornach: Rudolf Steiner Verlag 1976, pp. 150ff. Also in English as *Riddles of the Soul; Kunst und Kunsterkenntnis*, GA 271, 3rd edition, Dornach: Rudolf Steiner Verlag 1984, lectures of May 6, 1918 and June 1, 1918; *Anthroposophie – eine Zusammenfassung nach einundzwanzig Jahren*, GA 234, 6th edition, Dornach: Rudolf Steiner Verlag 1994, lectures of February 1 and 2, 1924. Also in English as *Anthroposophy and the Inner Life*, Rudolf Steiner Press.

113. Selg, op. cit. (note 2), pp. 308-310, 466-472, 515-518.

114. Steiner, *Von Seelenrätseln*, op. cit. (note 112). Also in English as *Riddles of the Soul*.

115. Steiner, *Kunst und Kunsterkenntnis*, op. cit (note 112), lecture of May 6, 1918, pp. 157ff, and June 1, 1918, pp. 177-179.

116. This experiment has been repeated multiple times with different student groups: A group of adults sings the melody of a familiar folk without words. When they are asked to sing it again to themselves, silently, most of the singers finish the inaudible song *at the same time*. When questioned, they say (and the experimenters could also observe) that they continued breathing exactly as they did during the "real" singing. Then the silent singing is repeated, but the singers are instructed to make a

conscious effort to breathe at intervals that are different from what they were accustomed to as they sang out loud. This time, the choir members do *not* finish the song together; that is, the song's temporal *gestalt* disintegrates.

117. A. Mikulski, E Herman, Die Hirnpulsation des Menschen, in: *Z. ges. Neurol. Psychiatr.* 1924 / 90, pp. 496-520; M. Resnikow, S. Dawidenkow, Beiträge aur Plethysmographie des menschlichen Gehirns, *Z. Neurol. Psychiatr.* 4 (1911), pp. 129-193; A. Mosso, *Über den Kresilauf des Blutes im menschlichen Gehirn*, Leipzig 1881.

118. F. Bender et al., Über das Verhalten des Liquordrucks bei psychischen Vorgängen, in: *Z. gest. Exp. Med.* 1951 / 117, pp. 349-358. Re: the history of research on the cerebrospinal fluid, see: R. M. Schmidt, *Der Liquor cerebrospinalis*, Stuttgart 1987, and Mikulski and Herman, op. cit. (note 117).

119. Sacks, op. cit. (note 33).

120. A. Damasio, *Decartes' Irrtum. Fühlen, Denken und das menschliche Gehirn*, Berlin 2006, and *Ich fühle, also bin ich*, Berlin 2007.

121. M. Spitzer, *Musik im Kopf*, Stuttgart 2004.

122. The same is true of the studies in P. Juslin, J. A. Sloboda (eds.), *Musik und Emotion*, 2003.

123. Spitzer, op. cit. (note 121), p. 39.

124. E. Kolisko, Physiologisches und Therapeutisches. Afterword to the book by V. Werbeck-Svärdstöm, *Die Schule der Stimmenthüllung*, 4th edition, Dornach: Verlag am Goetheanum 1984, pp. 161-173.

125. G. Husemann, Der Atemrhythmus des Gehirns im Liquor-Cerebrospinalis, in: *Sinnesleben, Seelenleben und Krankheitsbild*, op. cit. (note 35), pp. 44-65.

126. A. Husemann, *The Harmony of the Human Body*, op. cit. (note 66).

127. Rudolf Steiner points to the cultural and historical dimension of the transformation of respiratory consciousness in his lecture of November 30, 1919: *Die Sendung Michaels*, GA 194. Dornach: Rudolf Steiner Verlag 1994. Also in English as *The Mission of Michael*.

128. P. Mörsdorf, Weitgehende Schwerelosigkeit im Innern unseres Körpers und im Inneren der Körper der Tiere und Planzen. Manuscript, 1984; and Raumfahrt: Gut adaptiert, in: *Deutsches Ärzteblatt* 47, 11.20.1985, vol. 82.

129. P. Knoll, Über die Druckschwankungen in der Cerebrospinal-Flüssigkeit und den Wechsel in der Blutfülle des zentralen Nervensystems, in: *Sitz. Ber. Kais. Akad.Wiss.Wien. Math. Naturw. Classe* 93/94, B Abt. 3 (1886), pp. 217-148, as quoted in R. M. Schmidt, op. cit. (note 118), p. 59.

130. J. L. Shah, Positive lumbar extradural space pressure, in: *Br. J. Anesth.* 1994 / 73:pp. 309-314.

131. G. Schroth, K. Klose, Cerebrospinal fluid flow. Physiology of respiration-related pulsations, in: *Neuroradiology* 1992 / 35, pp. 10-15; P. Winkler, Cerebrospinal Fluid Dynamics in Infants Evaluated with Color Doppler US and Spectral Analysis: Respiratory versus Arterial Synchronization, in: *Radiology* 1994 / 192, pp. 423-430. R. R. Lee, R.A . Abraham, C. B. Quinn, Dynamic physiologic changes in lumbar CSF volume quantitatively measured by three-dimensional fast spin echo, in: *MRT Spine* 2001 / 26: pp. 1172-1178: "During normal respiration, the volume of the dural sac from T2 to the end of the dural sac is 28-42 mil. Under abdominal pressure, the lumbar intrathecal volume is reduced by up to 40 percent due to dilation of the epidural veins." As quoted in S. J. A. Friese, *Einfluss der Atmung auf die kranielle und spinale Liquorbewegung*, habilitation, U. Tübingen 2003, pp. 54 & 64.

132. See Winkler, ibid. (note 131)

133. R. Steiner, op. cit. (note 112).

134. In thoracic respiration, which develops around age ten, the direction of flow is supposedly reversed (Schroth and Klose, op. cit., note 141). "Thoracic respiration," however, is a secondary modification that is superimposed on primary, abdominal respiration.

135. S. J. A. Friese, op. cit. (note 131); U. Klose, C. Strik, C. Kiefer, W. Grodd, Detection of a Relation Between Respiration and CSF Pulsation with an Echoplanar Technique, in: *J. Magn. Res.Imaging* 2000 / 11: pp. 438-444.

136. D.I. J. Beentjes, The cochlear aqueduct and the pressure of cerebrospinal and endolabyrinthic fluids, *Acta Otolaryngologica* 73 (1972), pp. 112-120. As quoted in R. J. Marchbank, A. Reid, Cochlear and cerebrospinal fluid pressure: their interrelationship and control mechanisms, in: *British Journal of Audiology* 1990 / 24, pp. 17-187. Also important is a study by Q. Gopen et al., Anatomy of normal human cochlear aqueduct with functional implications, Hearing Research 107 (1997), pp. 9-22. The authors surmise that the perilymphatic duct (cochlear aqueduct) *insulates* the inner ear from cardiac pulsations and respiratory fluctuations in the CSF. This is the only indication I could find that considers the possibility of a functional interaction between the respiratory dynamic of the CSF and the perilymph of the ear.

137. In guinea pigs, an additional connection has been identified between the CSF system and the inner ear via the perineural sheath of the auditory

nerve: W. Arnold, C. von Illberg, Verbindungswege zwischen Liquor und Perilymphraum, in: *Arch. Lin. Exp. Ohr-, Nase- und Kehlk. Heilk.* 1971 / 198, pp. 247-261.

138. In my view, the question of the importance of thoracic and abdominal respiration for CSF movement remains open. Georg Soldner, who reviewed this chapter, suggests that with regard to developmental psychology, it would be important to investigate more exactly when and how the transition occurs from the infant's abdominal breathing to thoracic breathing and whether and how the constitutionally dominant type of respiration is reflected in CSF movement and/or in the condition of consciousness.

139. A musical physiology of CSF formation can be found in A. Husemann, *Der musikalische Bau des Menschen*, op. cit. (note 66), pp. 59-79. In English as *The Harmony of the Human Body.*

140. Schmidt-Thews, *Physiologie des Menschen*, 29th edition, Heidelberg 2004, a table of normal values in human physiology, inside of back cover.

141. R. Steiner, *Das Sonnenmysterium und das Mysterium von Tod und Auferstehung*, GA 211, 2nd edition, Dornach: Rudolf Steiner Verlag 1986, lecture of April 1, 1922, In English as *The Sun Mystery & the Mystery of Death and Resurrection*; Die Erkundung und Formulierung des Weltenwortes in der Ein- und Ausatmung. An anthroposophic discussion of venous and arterial circulation (A. Husemann, Studien zur Physiologie des Menschen, vol. 2) is currently in preparation.

142. Rudolf Steiner, *Anthroposophie. Eine Zusammenfassung nach einundzwanzig Jahren*, op. cit. (see note 112), 5th lecture (February 2, 1924).

143. Rudolf Steiner, *Geisteswissenschaftliche Impulse zur Entwickelung der Physik*, op. cit. (see note 68) vol. 1, pp. 125ff.

144. R. Steiner, *Theosophy. An Introduction to the Spiritual Processes in Human Life and in the Cosmos.* Anthroposophic Press 1994.

145. R. Steiner, *Grundlinien einer Erkenntnistheorie der Goetheschen Weltanschauung*, GA 2, 7th edition, Dornach: Rudolf Steiner Verlag 1979, p. 103, pp. 105-107. In English as *Goethe's Theory of Knowledge*, SteinerBooks 2008.

146. R. Steiner, *Anthroposophie als Kosmosophie – Zweiter Teil: Die Gestaltung des Menschen als Ergebnis kosmischer Wirkungen*, GA 208, 3rd edition, Dornach: Rudolf Steiner Verlag 1992, p. 87. In English as *Cosmosophy, vol. 2. Kunst und Kunsterkenntnis*, op. cit. (see note 112), lecture of June 1, 1918,

p. 172. See also Karl Trincher, *Wasser als Grundstruktur des Lebens und Denkens*, Vienna 1990; F. Teichmann, Strömen und Denken, in: *Elemente der Naturwissenschaft* 1973 / 18 (I): pp. 14-23. The creative nature of the etheric forces available in breathing, as metamorphoses of reproductive forces, is presented in the author's book *Der musikalische Bau des Menschen* (op. cit., note 66), pp. 79-84. In English as *The Harmony of the Human Body*.

147. R. Steiner, *Das Rätsel des Menschen. Die geistigen Hintergründe der menschlichen Geschichte*, op. cit. (see note 112), lecture of August 8, 1916, and *Kunst und Kunsterkenntnis*, op. cit. (see note 112), lectures of May 6, 1918 and June 1, 1918.

148. S. Duke-Elder, K. C. Wybar, *System of Ophthalmology*, vol. 2, St. Louis 1961, p. 286; Benninghoff, *Lehrbuch der Anatomie des Menschen*, vol. 3, 13th/14th edition, Munich/Vienna/Baltimore, p. 495.

149. The physical correlate of *musicality* remains unknown, although it must exist, since musicality is hereditary. The evidence presented here suggests the *width of the periplymphatic duct* and perhaps also that of the endolymphatic duct as a hypothesis that could be verified using magnetic resonance imaging or physical examination of pathological anatomical conditions. In musicians, the cross-section of one or both of these ducts should be larger than in non-musicians.

150. R. Steiner, *Die Erneuerung der pädagogsch-didaktischen Kunst*, 3rd edition, Dornach: Rudolf Steiner Verlag 1977, lecture of May 6, 1920.

151. R. Steiner, Die Psychologie der Künste, in: *Kunst und Kunsterkenntnis*, op. cit. (see note 112); A. Husemann, *Der musikalische Bau des Menschen* (op. cit. note 66), *Der Zahnwechsel des Kindes*, (op. cit., note 48), and Husemann (ed.) *Menschenwissenschaft durch Kunst* (op. cit, note 4).

152. See Schroth and Klose 1992, Winkler 1994 (note 131).

153. G. Husemann, *Sinnesleben, Seelenwesen und Krankheitsbild*, op. cit. (note 35).

154. To the extent that I can survey the situation, CSF-contacting neurons have been ascribed purely vegetative functions to date: B. Vigh, D. Vigh, Comparative Ultra Structure of the Cerebrospinal Fluid-Contacting Neurons, in: *International Review of Cytology* 1973 / 35, pp. 193-251. Academic Press, New York and London; B. Vigh, I. Vigh-Teichmann, Actual Problems of the Cerebrospinal Fluid-Contacting Neurons, in: *Microscopy Research and Technique* 1998 / 41: pp. 57-83; B. Vigh et al., The system of cerebrospinal fluid-contacting neurons. Its supposed role in the non-synaptic signal transmission of the brain, in: *Histology and Histopathology* 2004 / 19, pp. 607-628 (here p. 619).

155. Thanks to Claudia McKeen for drawing my attention to this painting.
156. Steiner, *Geisteswissenschaftliche Impulse zur Entwickelung der Physik*. Vol. 2, op. cit. (see note 68), lecture 14, March 14, 1920.
157. Steiner, *Kunst und Kunsterkenntnis*, op. cit. (see note 112), lecture of June 1, 1918. Steiner developed a "physiology of fantasy" that elaborates on the spiritual scientific thoughts presented by Schiller in his *Letters on the Esthetic Education of Man*. In physiological terms, respiration is the realm of free interplay between Schiller's "formal drive" and "material drive."
158. R. Steiner, *Entsprechungen zwischen Mikrokosmos und Makrokosmos*, GA 201, 2nd edition, Dornach: Rudolf Steiner Verlag 1987, lecture of April 23, 1920. In English as *Microcosm and Macrocosm*. See also G. Husemann, *Erdengebärde und Menschengestalt*, op. cit. (note 61).
159. R. Steiner, *Das Wesen des Musikalischen und das Tonerlebnis im Menschen*, op. cit. (note 21), first concluding remarks on December 29, 1920. In a different book, this author discusses Beethoven's liver cirrhosis from this perspective. (A. Husemann, *Der musikalische Bau des Menschen*, 4th edition, op. cit., note 66, pp. 159-163). See also the Afterword to this volume, pp. 94-97.
160. Steiner, *Das Wesen des Musikalischen und das Tonerlebnis im Menschen*, op. cit. (see note 21), lecture of March 7, 1923.
161. Steiner, *An Outline of Esoteric Science*, op. cit. (note 58)
162. Ibid., p. 207.
163. A. Husemann, *Der Zahnwechsel des Kindes*, op. cit. (note 48), and Der musikalische Bau des Menschen, op. cit. (note 66).
164. Steiner, *An Outline of Esoteric Science*, op. cit., pp. 182-183.
165. Ibid., p. 184.
166. Steiner, *Kunst und Kunsterkenntnis*, op. cit. (note 112), lecture of June 1, 1918.
167. See A. Husemann, Form und Zahl – zwei Tore ins Leben der Natur, in: A. Husemann (ed.), *Menschenwissenschaft durch Kunst*, op. cit (see note 4) pp. 179-185. (Lines 12 and 16 on page 182 should read "mammals" instead of "vertebrates.")
168. Steiner, *Kunst und Kunsterkenntnis*, op. cit. (see note 112), lecture of June 1, 1918.
169. G. and M. Glöckler, Das musikalische Geheimnis des platonischen Weltenjahres, in: *Das Goetheanum* 1995, pp. 729-730. Reprinted in A. Husemann (ed.), *Menschenwissenschaft durch Kunst*, op. cit. (note 4), pp. 186-194.

170. This confirms the musical physiology of CSF formation, which I developed (based on a sculptural exercise presented by Rudolf Steiner) in my book *Der musikalische Bau des Menschen* (op. cit., see note 66), long before I became aware of this connection. Also in English as *The Harmony of the Human Body*.

171. R. Steiner, Die Chymische Hochzeit des Christian Rosenkreutz, in: *Philosophie und Anthroposophie. Gesammelte Aufsätze 1904-1918*, GA 35, Dornach: Rudolf Steiner Verlag 1965, pp. 320-390.

172. H. Pfrogner, *Lebendige Tonwelt*, Munich: Langen-Müller-Verlag 1976.

173. See A. Husemann, *Der musikalische Bau des Menschen*, op. cit. (note 66), pp. 201ff. Also in English as *The Harmony of the Human Body*.

174. H. Ruland, *Ein Weg zur Erweiterung des Tonlebens*, Basel 1981.

175. See R. Steiner, *Das Wesen des Musikalischen und das Tonerlebnis im Menschen*, op. cit. (note 21), lecture of March 7, 1923. Also in English as *The Inner Nature of Music and the Experience of Tone*.

176. See A. Huseman, *Der musikalische Bau des Menschen*, op. cit. (note 66), pp. 205-208. Also in English as *The Harmony of the Human Body*.

177. Ibid.

178. H. Walwei-Wiegelmann (ed.), *Goethes Gedanken über Musik*, Frankfurt 1985, p. 197.

179. J. M. Hauer, *Vom Melos zur Pauke*, Vienna, undated, Universal-Edition no. 8395.

For the latest information and a complete listing of SteinerBooks,
visit our website at:

www.steinerbooks.org

www.ingramcontent.com/pod-product-compliance
Lightning Source LLC
Chambersburg PA
CBHW030118170426
43198CB00009B/659